THE SCIENCE OF ICE CREAM

RSC Paperbacks

RSC Paperbacks are a series of inexpensive texts suitable for teachers and students and give a clear, readable introduction to selected topics in chemistry. They should also appeal to the general chemist. For further information on all available titles contact:

Sales and Customer Care Department, Royal Society of Chemistry, Thomas Graham House, Science Park, Milton Road, Cambridge CB4 0WF, UK Telephone: +44 (0)1223 432360; Fax: +44 (0)1223 426017; E-mail: sales@rsc.org

Recent Titles Available

Future titles may be obtained immediately on publication by placing a standing order for RSC Paperbacks. Information on this is available from the address above.

RSC Paperbacks

THE SCIENCE OF ICE CREAM

CHRIS CLARKE
Unilever R&D Colworth, Sharnbrook, Bedfordshire

RS•C
advancing the chemical sciences

ISBN 0-85404-629-1

A catalogue record for this book is available from the British Library

Published by The Royal Society of Chemistry,
Thomas Graham House, Science Park, Milton Road,
Cambridge CB4 0WF, UK

Registered Charity Number 207890

For further information see our web site at www.rsc.org

Typeset by H Charlesworth & Co Ltd, Huddersfield, UK
Printed by TJ International Ltd, Padstow, Cornwall, UK

Preface

The almost invariable response when people find out that I am an ice-cream scientist is: 'What a great job! But is there science in ice cream? Do you invent new flavours?' As a physicist I do not invent new flavours, but my job does involve, amongst other things, inventing new textures. To explain this I normally briefly describe what ice cream is made of and how this relates to the texture you experience when you eat it. Fortunately, almost everybody likes ice cream, so, unlike most other areas of research in which I have worked, I usually manage to finish explaining the science before the listener's eyes glaze over. In fact ice cream is a great vehicle for talking about science. My colleagues and I regularly receive invitations to speak to schools, science societies and at events such as the Cambridge Science Festival and National Science Week. These talks have proved popular (I hope not only because of the free samples), and as a result I have received requests to write articles for journals such as *Education in Chemistry* and *Physics Education*, and to help with material for the Royal Institution Christmas Lectures. So when the RSC asked me whether I thought there would be a market for a book on *The Science of Ice Cream*, I felt confident enough that there would be to agree to write one.

The aim of this book is to show that there is a great deal of science in ice cream, and in particular to demonstrate the link between the microscopic structure and the macroscopic properties. It is naturally biased towards physics, physical chemistry and materials science as these are the areas in which I trained. The book is aimed at schools and universities, and a scientific background is required to understand the more technical sections. I have attempted to make it readable by 16–18-year-olds, and many sections are suitable for adaptation by GCSE science teachers. I have unashamedly made reference to the giants of chemistry and physics such as Newton, Einstein, Boyle, Gibbs, Kelvin, Laplace and Young where the laws and equations that bear their names are relevant. I hope that as a result teachers reading this book will find in ice cream useful illustrations of a number of scientific principles. Some sections could be used with younger pupils, especially Chapter 8, which describes experiments on ice cream that can mostly be performed without specialist equipment in a classroom or kitchen. This book should also provide a useful introduction to ice cream for someone who has recently joined the food industry but I nonetheless hope that it will be accessible to any interested reader who is prepared to skip the most technical sections. I have included a glossary to explain the technical terminology.

I could not have written this book without the assistance of a large number of people. Firstly I would like to thank Unilever for permission to publish it. My colleagues at Unilever Colworth have provided data, images and suggestions. I am especially grateful to Laurie Bender, Allan Bramley, Deryck Cebula, Bruno Chavez, Andrew Cox, Paul Doehren, Viki Evans, Dudley Ferdinando, Dick Franklin, Andy Hoddle, Martin Izzard, Danny Keenan, Mark Kirkland, Linda Jamieson, Dan Jarvis, Jean-Yves Mugnier, Tricia Quail, Andrew Russell, Susie Turan, and Paul Trusty. Javier Aldazabal from CEIT, San Sebastian, Spain, WCB Ice Cream, and the London Canal Museum kindly supplied images. Elsevier Science, IOP Publishing Ltd, Microscopy and Analysis and The Royal Society of Chemistry gave their permission to reproduce previously published material. Finally, I would like to thank my wife Alexandra for reading the draft, my mother Lorrie and my niece Charlotte for trying out the experiments and my son Theo, whose arrival nearly provided sufficient impetus for me to finish the book!

Table of Contents

Chapter 7
Ice Cream: A Complex Composite Material   135

Chapter 8
Experiments with Ice Cream and Ice Cream Products   166

Glossary

Ageing: A step in the manufacturing process in which pasteurized, homogenized mix is held at about 4 °C for several hours, during which some of the fat crystallizes and some of the protein coating on the fat droplets is replaced by emulsifiers.

Attribute: A term for one aspect of the sensory properties, for example firmness, smoothness, iciness *etc.*, used in analytical sensory measurements.

Coarsening: The increase in the mean size and reduction in number of particles in a colloidal dispersion at constant total volume, thereby lowering the energy.

Composite material: A material obtained by combining two or more component materials on a microscopic or macroscopic level (*i.e.* not at the molecular level). The components do not dissolve in each other, and the interfaces between them can be identified.

Contiguity: A measure of the connectivity of one phase in a composite material.

Couverture: Chocolate analogue that is made with fats other than cocoa butter, for example coconut oil. Couvertures have a wider range of textures than chocolate and can be flavoured, for example, with lemon, strawberry or yoghurt.

Dasher: A mixing device that rotates inside the barrel of a scraped surface heat exchanger and to which scraper blades are attached. The dasher has two functions: to scrape ice crystals off the barrel wall and to shear the ice cream as air is injected thereby breaking up large air bubbles and mixing in the ice crystals. Dashers may be open (*i.e.* they occupy a small proportion of volume of the barrel, typically 20–30%)

or closed (*i.e.* they occupy a large proportion of the volume, typically 80%).

Destabilized (de-emulsified) fat: Fat that has undergone partial (or total) coalescence so that it is no longer in the form of a fine emulsion.

Dextrose equivalent (DE): A measure of the extent to which the polysaccharides have been broken down into smaller molecules in corn syrups. The higher the DE, the lower the average molecular weight. Dextrose has a DE of 100 and starch has a DE of 0.

Eutectic mixture: The specific mixture of two compounds that has the lowest melting point of any such mixture. Eutectic mixtures (unlike other mixtures) melt and freeze at a constant temperature, called the eutectic temperature.

Factory freezer: A scraped surface heat exchanger in which the first stage of ice cream freezing takes place.

Failure mechanism: The manner in which a material breaks (fails) when it is deformed. Water ices typically undergo rapid, brittle failure, whereas ice cream typically undergoes more gentle plastic failure.

Freeze-concentration: The process by which a solution becomes more concentrated as it is frozen. The ice that is formed excludes the solute molecules. Thus as freezing proceeds, the number of water molecules in the solution decreases, but the number of solute molecules does not, so it becomes more concentrated.

Glass transition: When, for example, a sucrose solution is cooled down it becomes more concentrated due to freeze-concentration. Since the sucrose does not crystallize easily the solution becomes very viscous and, as it is cooled, the molecular motion becomes very slow. Eventually the molecular motion effectively stops and the viscosity becomes so large that the solution effectively becomes a solid. However, unlike a crystal, the molecules are not ordered on a lattice, but have a disordered liquid-like structure, known as a glass. The change to a glassy solid is known as the glass transition.

Hardening: The second freezing step in the manufacturing process in which partly frozen ice cream from the factory freezer is placed in a very cold environment in order to cool it rapidly to a temperature at which coarsening of ice crystals and air bubbles is halted and the ice cream is hard enough for further processing, such as dipping in chocolate.

Hydrogen bonding: A strong inter- or intramolecular attraction that occurs between hydrogen and oxygen atoms.

Inclusions: Pieces of fruit, nuts, chocolate, biscuit, cookie dough, marshmallow, toffee *etc.* that can be mixed into ice cream.

Matrix: The continuous phase in ice cream and water ice. It is a viscous solution of sugars (and, depending on the formulation, other ingredients such as polysaccharides, milk proteins, colours, flavours etc.). It acts as a glue, holding the ice crystals, air bubbles and fat droplets together.

Meltdown: An empirical measure of the rate at which ice cream melts when exposed to warm temperatures, usually determined by measuring the amount of melted ice cream that drips through a wire mesh as a function of time in a temperature-controlled environment.

Milk solids non fat: All of the components of milk other than water and fat, *i.e.* protein, lactose, vitamins, minerals and other minor constituents.

Mono-/ di-/ tri-glyceride: A molecule consisting of one/ two/ three fatty acids esterified to a glycerol molecule. Mono- and diglycerides are emulsifiers, triglycerides are fats.

Mono-/ di-/ tri-/ oligo-/ polysaccharide: A molecule consisting of one/ two/ three/ several/ many saccharide units. Monosaccharides are the simplest sugars, and conform to the chemical formula $(CH_2O)_n$.

Overrun: A measure of the amount of air in ice cream defined by (volume of ice cream −volume of mix)/volume of mix, expressed as a percentage.

Partial coalescence: When two fat droplets that contain some solid and some liquid fat coalesce they form a cluster that retains some of the original droplets' individual nature. Thus they coalesce, but only partially.

Percolation: A microstructure in which a continuous path can be traced from one side of the material to the other in a single phase is said to be percolated; for example, the ice crystals in a high ice content water ice.

Polyelectrolyte: A polymer that dissociates on dissolving in water to give a multiply charged polymer and an equivalent amount of ions of small charge and opposite sign.

Propagation: The growth of ice crystals as the temperature is lowered during hardening, accompanied by an increase in the total ice phase volume.

Quiescent freezing: Freezing without agitation, used for example in the production of some moulded water ices.

Recrystallization: The coarsening of ice crystals, *i.e.* the increase in mean size at constant ice phase volume.

Residence time: The length of time that ice cream mix spends inside the barrel of the factory freezer.

Rheology: The study of the deformation and flow of liquid and semi-solid materials.

Scraped surface heat exchanger: A class of equipment designed to remove heat from viscous liquids. Scraped surface heat exchangers normally consist of a cylindrical barrel, the outside of which is cooled. The inside of the barrel is scraped to remove solidifying material and thereby increase the heat flow.

Slush freezing: Freezing with agitation, for example in a factory freezer.

Wheying off: The phase separation of milk proteins and stabilizers in the matrix.

SYMBOLS

a	absorbance
A	area
B	breadth

b	constant in power law fluid
c	heat capacity
C	contiguity
d	displacement
D	depth
E	energy
F	force
G	shear modulus
G'	storage modulus
G''	loss modulus
g	gravity
H	hardness
h	height
K	cryoscopic constant
l	length
L	latent heat
m	mass
n	number
p	pressure
Q	heat
R	gas constant
r	radius
s	solubility
S	span
t	time
T	temperature
v	velocity
V	volume
x	molality
Y	Young's modulus
ε	strain
η	viscosity
ϕ	volume fraction
$\dot{\gamma}$	shear rate
γ	surface tension
σ	shear stress
ρ	density

Chapter 1

The Story of Ice Cream

WHAT IS ICE CREAM?

Ice cream is an enormously popular food. The term 'ice cream' in its broadest sense covers a wide range of different types of frozen dessert. The main ones are

- dairy ice cream – a frozen, aerated mixture of dairy ingredients, sugars and flavours.
- non-dairy ice cream – made with milk proteins and vegetable fat.
- gelato – an Italian-style custard-based ice cream that contains egg yolks.
- frozen yoghurt – which may contain lactic acid organisms, or simply yoghurt flavour.
- milk ice – similar to ice cream, but unaerated and containing less dairy fat.
- sorbet – fruit based, aerated sugar syrup that contains neither fat nor milk.
- sherbet – similar to a sorbet, but containing some milk or cream.
- water ice – frozen sugar syrup with flavour and colour, such as an 'ice lolly'.
- fruit ice – similar to water ice, but made with real fruit juice.

What these all have in common is that they are sweet, flavoured, contain ice and, unlike any other frozen food, are normally eaten in the frozen state.

The legal definition of ice cream varies from country to country. In the UK 'ice cream' is defined as a frozen food product containing a minimum of 5% fat and 7.5% milk solids other than fat (*i.e.* protein, sugars and minerals), which is obtained by heat-treating and

1

subsequently freezing an emulsion of fat, milk solids and sugar (or sweetener), with or without other substances. 'Dairy ice cream' must in addition contain no fat other than milk fat, with the exception of fat that is present in another ingredient, for example egg, flavouring, or emulsifier.[1] In the USA, ice cream must contain at least 10% milk fat and 20% total milk solids, and must weigh a minimum of $0.54\,kg\,l^{-1}$. Until 1997, it was not permitted to call a product 'ice cream' in the USA if it contained vegetable fat.

Ice cream is often categorized as premium, standard or economy. Premium ice cream is generally made from best quality ingredients and has a relatively high amount of dairy fat and a low amount of air (hence it is relatively expensive), whereas economy ice cream is made from cheaper ingredients (*e.g.* vegetable fat) and contains more air. However, these terms have no legal standing within the UK market, and one manufacturer's economy ice cream may be similar to a standard ice cream from another.

Most people are very familiar with the appearance, taste and texture of ice cream and there are many recipes for making it in cookery books. However, few people know *why* certain ingredients and a time-consuming preparation process are required. The answer is that ice cream is an extremely complex, intricate and delicate substance. In fact, it has been called "just about the most complex food colloid of all".[2] The science of ice cream consists of understanding its ingredients, processing, microstructure and texture, and, crucially, the links between them. This requires a whole range of scientific disciplines, including physical chemistry, food science, colloid science, chemical engineering, microscopy, materials science and consumer science (Figure 1.1).

The ingredients and processing create the microstructure, which is shown schematically in Figure 1.2. It consists of ice crystals, air bubbles and fat droplets in the size range 1 μm to 0.1 mm and a viscous solution of sugars, polysaccharides and milk proteins, known as the matrix. The texture we perceive when we eat ice cream is the sensory manifestation of the microstructure. Thus, microstructure is at the heart of the science of ice cream, and forms the central theme running through this book.

To describe the science of ice cream, it is first necessary to describe some of the physical chemistry and colloid science that underpins it; these are laid out in Chapter 2. Chapters 3 and 4 cover the ingredients and the ice cream making process respectively. Chapter 5 focuses on the production of various types of ice cream product. The physical and sensory measurements used to quantify and describe it are discussed in Chapter 6, and the microstructure, and its relationship to the texture, is examined in Chapter 7. Finally, Chapter 8 describes a number of

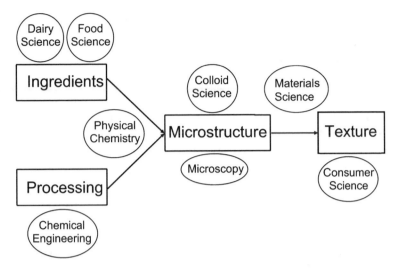

Figure 1.1 *The sciences of ice cream*

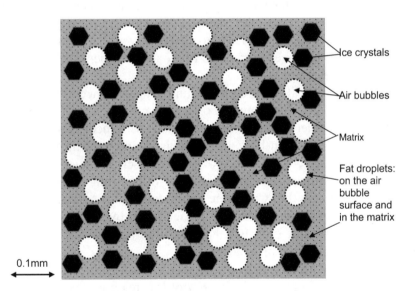

Figure 1.2 *Schematic diagram of the microstructure of ice cream*

experiments that illustrate the science of ice cream, which may be performed in the laboratory, classroom or kitchen. We begin, however, by looking at where and when ice cream was invented, and how it has evolved into the huge range of products eaten by billions of people all around the world today.

THE HISTORY OF ICE CREAM

Ice cream as we recognize it today has been in existence for at least 300 years, though its origins probably date much further back. The history of ice cream is full of myths and stories, which have little real evidence to support them. A typical 'history' begins with the Roman Emperor Nero (AD 37–68) who is said to have eaten fruit chilled with snow brought down from the mountains by slaves. Elsewhere, Mongolian horsemen are reputed to have invented ice cream. They took cream in containers made from animal intestines as provisions on long journeys across the Gobi desert in winter. As they galloped, the cream was vigorously shaken, while the sub-zero temperature caused it to freeze simultaneously. The expansion of the Mongol empire spread this idea through China, from where Marco Polo reputedly brought the idea to Italy when he returned from his travels in 1296. It has been claimed that ice cream was introduced to France from Italy when the 14-year-old Catherine de Medici was married to the Duc d'Orleans (later Henri II of France) in 1533. Her entourage included Italian chefs who brought the recipe for ice cream with them. The secret of making ice cream remained known to only a few. So precious was it that Charles I of England is said to have offered his French chef a pension of £500 per year to keep his recipe secret.

However, historical research has found little evidence to support any of these stories. The only mention of ice in connection with Nero comes from Pliny the Elder in the first century AD, who records the discovery that water that has been boiled freezes faster and is healthier. There is no mention of ice cream in any of the manuscripts describing Marco Polo's travels. Indeed, modern historians doubt that he even reached China. It is unlikely that Catherine's chefs knew how to make ice cream since, at that time, the method of refrigeration by mixing ice and salt was known in Europe only to a handful of scientists. Nor is there any documentary evidence for Charles' chef.

We cannot be absolutely sure of exactly who invented ice cream, or where and when. In reality, the history of ice cream is closely associated with the development of refrigeration techniques and can be traced in several stages related to this.

1. Cooling food and drink by mixing it with snow or ice.
2. The discovery that dissolving salts in water produces cooling.
3. The discovery (and spread of knowledge) that mixing salts and snow or ice cools even further.
4. The invention of the ice cream maker in the mid-19th century.

5. The development of mechanical refrigeration in the late 19th and early 20th centuries.

Concurrent with these a vast range of recipes have been developed, spanning the spectrum from chilled fruit juices to what we understand as ice cream today. The evolution of ice cream has been described in detail in the two excellent histories listed in the Further Reading and the following summary owes much to them.

Ice has been used to chill food and drink for at least 4000 years in many different parts of the world. Ice cellars dating back to 2000 BC have been discovered in Mesopotamia (present day Iraq). Records from the Zhou dynasty in China, *ca.* the 11th century BC, describe the role of the court iceman, who had a large staff responsible for harvesting ice every winter and storing it in cellars to be served with drinks in the summer. An ice-cooled dessert, made from water buffaloes' milk mixed with flour and camphor is recorded in the Tang dynasty (AD 618–907). In Greece, drinks were served with snow in about 500 BC and a Roman cookery book dating from the first century AD includes recipes for sweet desserts that are sprinkled with snow before serving. Sweet drinks cooled with ice are recorded in Persia in the second century AD. Rather than harvesting ice during the winter, the Persians exploited the cold desert nights to freeze water that had been placed in shallow pits.

The first significant step forward was the discovery that water is cooled when salts are dissolved in it, such as common salt (sodium chloride), saltpetre (potassium nitrate), sal-ammoniac (ammonium chloride) or alum (a mixture of aluminium sulfate and potassium sulfate). When the crystals dissolve, the strong bonds between the ions are broken, extracting heat from the surrounding water, so the temperature drops. Adding a mixture of 5 parts ammonium chloride and 5 parts potassium nitrate to 16 parts water at 10 °C causes the temperature of the mixture to drop to about − 12 °C, sufficient to freeze a vessel of pure water immersed in it. This phenomenon was first recorded in an Indian poem from the fourth century AD, and described in detail in an Arabic medical textbook from 1242. Another book in Arabic, containing sorbet recipes, appeared at about the same time.

These methods were known in the West by the early 16th century. Soon thereafter, in 1589, Giambattista Della Porta, a scientist from Naples made the breakthrough that paved the way for making ice cream as we know it today. He reported the discovery, possibly based on Arabic knowledge of cooling techniques, that much greater cooling could be achieved by mixing ice and salt. (This effect is explained in Chapter 2.) Della Porta used this discovery to freeze wine in a glass placed in a mixture of ice and salt. Others became aware of the phenomenon, and the first reports of iced desserts produced using this method of freezing began to appear in the 1620s. These appear to have been water ices, rather than ice cream. Water ices were served at banquets in Paris, Naples, Florence and Spain during the 1660s.

The earliest evidence for ice cream in England comes from a list of the food that was served at the feast of St George at Windsor in May 1671; ice cream was served, but only at King Charles II's table. Techniques and recipes then developed, notably in France. In 1674 Nicholas Lemery published a recipe for water ice, and two years later Pierre Barra described freezing a mixture of fruit, cream and sugar by using snow and saltpetre. L. Audiger noted the importance of whipping the mixture during freezing to break up large ice crystals in 1692. This technique still forms the basis of ice cream making today.

Water ices and ice creams became a luxury served by the aristocracy at their banquets. More recipe books appeared in the 18th century, which helped to widen knowledge of how to make them. This was accompanied by the production of special moulds, ice pails (for keeping ices cold) and serving cups. The extent of the spread of ice cream can be seen by the orders for the utensils associated with its production and serving. Those who are recorded as having such services include Louis XV of France, Gustaf III of Sweden and Catherine the Great of Russia. Ice cream remained the preserve of royalty and aristocracy in Europe, and of high-ranking officials in the USA, where in 1744 the governor of Maryland was one of the first people recorded to have served it. A Presidential connection with ice cream began with George Washington, who often served it at official functions, as did the third president, Thomas Jefferson. After his time as American envoy in France, Jefferson brought back an ice cream recipe from his French chef. Ice cream was served at the inaugural dinner for the fourth President, James Madison, and subsequently became a regular feature of the White House menu.

By the beginning of the 19th century, ice cream had started to move from the tables of the aristocracy into restaurants and cafes that served the well-off middle classes. This was accompanied by an increasing interest in good food, reflected in the appearance of many books on cooking. Ice cream was made by hand, by placing a bowl containing ice cream mix in a barrel filled with ice and salt, and using a scraper to remove growing ice crystals from the sides of the bowl, until in the 1840s Nancy Johnson of Philadelphia invented the first ice cream making machine. Her invention consisted of two spatulas that fitted tightly into a long cylindrical barrel. The spatulas contained holes and were attached to a shaft that could be rotated with a crank. The outside of the cylinder was cooled with a mixture of salt and ice. The holes made it easier to rotate the spatulas in the mix, while scraping ice crystals off the inner wall of the cylinder. This invention simplified ice cream production and ensured a more uniform texture than had previously been possible. At about the same time, people in Europe were coming up with other ideas for ice cream making machines and many patents were published both in Europe and the USA in the following decades. Mechanization meant that ice cream could be produced more cheaply and in much larger volumes. Jacob Fussell, a dairy farmer from Baltimore, USA, is commonly considered to be the founder of the modern ice cream industry. Faced with a surplus of milk during the summer, he decided to sell it as ice cream. He built the first ice cream factory in 1851, and subsequently expanded to Washington, Boston and New York, where he sold ice cream at a price that ordinary people could afford.

In England, ice cream became available to the masses towards the end of the 19th century. This was due in part to the emigration of Italians, many of whom became ice cream vendors in cities around the world. Of the approximately ten thousand Italians living in England and Wales at the time of the 1890 census, just under one thousand listed their occupation as street vendors, most of whom sold ice cream. They sold it in small, thick walled glasses, known as 'penny-licks'. These were usually wiped with a cloth and re-used, and were thus a considerable health hazard, particularly for children. The street vendors would drum up business by calling out 'ecco un poco', Italian for 'try a little'. The words 'ecco un poco' became 'hokey pokey', now meaning either poor quality ice cream, or deception/trickery.[3] Poor hygiene standards necessitated the introduction of regulations around the turn of the century.

One person who was not impressed with the quality of street-sold ice cream was Agnes Marshall (1855–1905). She was a celebrity cook with an interest in new technology. As well as inventing an ice cream maker (Figure 1.3) she wrote several books, including two dedicated to ice cream. She toured extensively, lecturing and demonstrating her techniques to large audiences and campaigned for better standards of food hygiene. She can also claim to have invented the ice cream cone, in 1888, since a recipe for 'cornets with cream' appears in one of her books. However, the cone really took off at the 1904 St Louis World's Fair, when a stall selling ice cream ran out of dishes in which to serve it. Ernest A. Hamwi, a waffle vendor in the next-door stall, had the bright idea of rolling up his waffles as cones instead. There were tens of

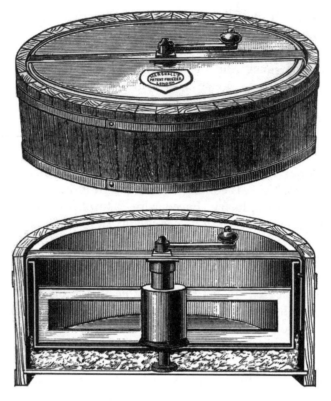

Figure 1.3 *Mrs Marshall's Ice Cream maker*
(Courtesy of the London Canal Museum, www.canalmuseum.org.uk)

thousands of visitors at the Fair, so the idea then spread rapidly throughout the USA.

Ice cream making required a source of ice. Starting in the first half of the 19th century, ice was collected from rivers and lakes in Norway, Sweden, Canada and the northern USA and then shipped to cities in Europe, the USA and even as far away as Calcutta! However, physicists and engineers were developing techniques for artificial refrigeration based on liquefying gases, such as propane and ammonia. The gas is compressed until it liquefies, during which it heats up. It is then cooled to near ambient temperature whilst still pressurized. Finally the pressure is released, which allows the liquid to expand and evaporate, extracting heat from the surroundings in the process, hence providing refrigeration (Experiment 1 in Chapter 8 demonstrates this principle). Several cooling machines were invented, but Carl von Linde's invention, demonstrated at the 1873 World's Fair in Vienna, was the first really successful one. Using ammonia gas in a closed circuit, he could rapidly produce substantial quantities of ice. Eventually, artificial production took over from harvesting as the main source of ice.

In the first few decades of the 20th century ice cream production expanded and became industrialized. The major reason for this was technological developments in the production and transport of ice cream. The development of mechanical refrigeration allowed the use of chilled brine (a concentrated salt solution that freezes well below 0 °C) as the refrigerant instead of salt and ice. This greatly increased the rate of heat transfer between the ice cream mix and the refrigerant, and hence the speed of production. In 1927 Clarence Vogt invented the continuous freezer, with a horizontal (rather than a vertical) cylinder. Mix was pumped in at one end and ice cream pumped out from the other, so that ice cream could be made continuously, rather than in batches. The modern ice cream factory freezer had arrived. (Note the distinction between the freezer in the factory, which converts ice cream mix into frozen ice cream, and the domestic freezer in which you keep it cold before consumption. Throughout this book the term 'factory freezer' is used for the former and 'freezer' for the latter.) These developments were accompanied by the introduction of pasteurization, which reduced concerns over the safety of ice cream, and homogenization, which produced a smoother, creamier product by breaking the fat into tiny droplets. Finally, better transport through the railways and the automobile made the supply of ingredients and the distribution of products possible over much greater distances than before.

Many of the names familiar to consumers around the world today have their origins in this period. In the USA, Breyers, which had consisted of a number of shops in the 1870s, built new factories and expanded its annual production of ice cream to nearly four million litres by 1914. In the UK, Wall's set up its first ice cream factory in Acton, London in 1922. Ice cream was sold from tricycles, and the phrase 'Stop me and buy one' became very familiar to ice cream consumers. Production came to a halt during the Second World War due to shortages of ingredients and the need to convert the factories to produce essential foods, such as margarine. When production resumed after the war, rationing was still in place, and it was forbidden to use cream to make ice cream. Manufacturers therefore switched to using vegetable fats and milk powder. By the time rationing ended in 1953, the British public had become accustomed to the taste of ice cream produced with vegetable fat. For this reason, and also because it is cheaper, a substantial amount of ice cream in the UK is still made with vegetable fat today. Partly as a result of a shortage of manpower after the war, the main sales outlet changed from tricycles to freezers in corner shops. As their businesses grew, large companies such as Wall's, which became part of Unilever in 1929, and Lyons Maid, which was bought by Nestlé in 1992, set up research and development departments to study ice cream and its manufacture. One of the scientists who conducted research for Lyons in the 1950s was a young chemist, Margaret Roberts, who later became better known under her married name of Thatcher. In recent years many other companies have joined the long-established manufacturers. Two of the best known of these are Häagen Dazs and Ben and Jerry's. Reuben Mattus founded Häagen Dazs in New York in 1960. He chose the (meaningless) name because it sounded Danish and was therefore associated with dairy produce. This fitted well with his new product, a high quality, high price ice cream. Ben Cohen and Jerry Greenfield set up an ice cream parlour in an abandoned petrol station in Burlington, Vermont in 1978. Their all-natural ice cream company with a strong social mission became famous after winning a major court battle with Häagen Dazs (owned at the time by Pillsbury). Ben and Jerry's has recently been acquired by Unilever and is developing into a world-wide business.

THE GLOBAL ICE CREAM MARKET

Ice cream is made and eaten in almost every country in the world. The total worldwide production of ice cream and related frozen desserts was 14.4 billion litres (l) in 2001, *i.e.* an average of 2.4 litres per person, worth £35 billion.[4] Unilever and Nestlé are the largest worldwide producers, with about 17 and 12% of the market respectively. A huge range of different flavours is available, including savoury ones. Differences in culture and climate produce wide variations in the amounts, types and flavours of ice cream produced and consumed in different countries.

The USA is the largest producer of ice cream (about 6 billion l per annum) and has a per capita annual consumption of about 22 l; only New Zealanders eat more, with an average consumption of 26 l. Some 9% of all the milk produced in USA is used to make ice cream, and more than 90% of US households buy it. It is often eaten as a snack, much as biscuits are eaten in the UK. Sales of ice cream in the US in 2000 were about $20 billion (£13 billion). Approximately two-thirds of this was sold in scoop shops, restaurants, retail outlets *etc.* and eaten out of the home. One-third was sold in supermarkets, grocery shops *etc.*, mostly as half-gallon (2.2 l) tubs. More than half of the sales were premium ice cream; low-fat ice cream, frozen yoghurt, and sherbet account for smaller ($< 10\%$) but significant proportions of the market. Vanilla is the most popular flavour, accounting for about a quarter of sales, followed by chocolate. Ice cream with pieces of other components (known as inclusions), such as cookie dough, marshmallows, fruit chunks, nuts, chocolate, toffee or fudge, is becoming increasingly popular, and now accounts for nearly a quarter of sales.

The European per capita ice cream consumption figures are surprising at first sight. One might expect that more ice cream would be consumed in hot southern European countries such as Spain (about 6 litres per person per year) and Portugal (4 l) than in cold northern European countries such as Sweden (12 l) and Germany (8 l). However, the reverse is true. The main reason for this is that northern Europeans are used to consuming lots of milk, cheese, butter *etc.* whereas the southern European diet contains much less dairy produce. Another factor that influences consumption is whether households own a large freezer in which to keep quantities of ice cream. This in turn may be influenced by local building regulations! The exception to this north–south divide is Italy (9 l), where there is a great tradition of making and eating ice cream. In many European countries bars and stick products are more important than tubs.

The UK falls roughly in the middle of the list of European countries, with a per capita consumption of about 7 l, and annual sales of about £1.5 billion. A small number of large companies, such as Wall's (Unilever), Mars and Richmond Foods (which produces ice cream for Nestlé, and several supermarkets' own brands) have substantial market shares, but about half is taken by the several hundred small independent companies. These mostly employ fewer than ten people, and sell only locally. The most popular flavours in the UK are vanilla, chocolate and strawberry, and, like the US, ice cream often contains inclusions to provide greater interest and variety for the consumer. *Magnum* is the largest single brand – 41% of adults in the UK have bought one.[5]

In other parts of the world, the market is very different. For example, in southeast Asia the largest demand is for refreshing products, such as water ices. Ice cream comes in flavours that seem very strange and exotic to Western palates – for example green tea and red bean ice cream in Japan, sweet corn ice cream in Malaysia, chilli ice cream in Indonesia and sesame seed ice cream in Korea.

SELLING ICE CREAM: FUN, INDULGENCE AND REFRESHMENT

Two factors that have a major effect on the sales of ice cream products are the weather and advertising. Ice cream sales in the UK are very seasonal, peaking in the summer. In France, 65% of sales are made between June and September, and in Italy the average consumption per capita per month is 0.1 l in January and 1.3 l in July. The weather can have a substantial impact on sales, especially at particular times, such as holiday weekends. While this is beyond anyone's control, the large companies try to boost their sales by spending several million pounds each year on promoting their products. Ice cream advertisements and television commercials are often based on one of three themes: fun, indulgence or refreshment. A fun image is typically used to promote products aimed at children or families. The advertising for premium products (which may well use an element of sexual attraction) usually aims to project an indulgent image. Water ice products, for example ice lollies, are frequently marketed on their ability to cool down and refresh the consumer. Whether a product is indulgent or refreshing largely depends on its ingredients and the processing method by which it is produced. We will look at these in Chapters 3–5.

REFERENCES

1. 'The Food Labelling Regulations 1996', Statutory Instrument 1499, http://www.hmso.gov.uk.
2. E. Dickinson, 'An Introduction to Food Colloids', Oxford University Press, Oxford, 1992.
3. 'New Shorter Oxford English Dictionary', Clarendon Press, Oxford, 1993.
4. Global and US Market Information from the International Dairy Foods Association (www.idfa.org), and *Dairy Ind. Int.*, 2002, **67**, 27.
5. UK Market Information from 'Ice Creams and Frozen Desserts: Key Note Market Report Plus' 7th Edition, ed. E. Clarke, Hampton, 2000, and M. Stogo, 'Ice Cream and Frozen Deserts: a Commercial Guide to Production and Marketing', J. Wiley, New York, 1998.

FURTHER READING

C. Liddell and R. Weir, 'Ices: The Definitive Guide', Grub Street, London, 1995.
P. Reinders, 'Licks, Sticks and Bricks – A World History of Ice Cream', Unilever, Rotterdam, 1999.

Chapter 2

Colloidal Dispersions, Freezing and Rheology

A typical ice cream consists of about 30% ice, 50% air, 5% fat and 15% matrix (sugar solution) by volume. It therefore contains all three states of matter: solid ice and fat, liquid sugar solution and gas. The solid and gas are small particles – ice crystals, fat droplets and air bubbles – in a continuous phase, the matrix. To understand the creation of the microstructure during the manufacturing process we must first introduce some concepts from the physical chemistry of colloids, freezing and rheology (the study of the deformation and flow of materials).

COLLOIDAL DISPERSIONS

Colloidal dispersions consist of small particles of one phase (solid, liquid or gas) in another continuous phase. The particle size may range from nanometres to tens of microns. There are eight different types of colloidal dispersion, summarized in Table 2.1.

Colloidal dispersions have a very large surface area for their volume. Therefore the surface properties of the phases have a large influence on the properties as a whole. Ice cream is simultaneously an emulsion (fat droplets), a sol (ice crystals) and a foam (air bubbles), and also contains other colloids in the form of casein micelles, other proteins and polysaccharides in the matrix.

Emulsions

Emulsions are dispersions of droplets of one liquid in another. Many foods, for example mayonnaise, vinaigrette salad dressing, milk, cream and ice cream are oil in water emulsions, *i.e.* the oil is dispersed as droplets in a continuous aqueous phase. Low-fat margarines are

Table 2.1 *Classification of colloidal dispersions*

Continuous phase	Dispersed phase	Name	Examples
Solid	Solid	Solid sol	Ruby, glass, composites, ceramics, bone
Solid	Liquid	Solid emulsion	Bitumen, asphalt, opal, pearl, jelly
Solid	Gas	Solid foam	Expanded polystyrene, pumice
Liquid	Solid	Sol	Ink, paint, blood, toothpaste, mud
Liquid	Liquid	Emulsion	Milk, mayonnaise, cream
Liquid	Gas	Foam	Head on beer, bubble bath
Gas	Solid	Aerosol	Smoke, dust
Gas	Liquid	Aerosol	Mist, fog, clouds, deodorant

water-in-oil emulsions, *i.e.* the water is dispersed in a continuous oil phase.

Liquids behave as if they have an elastic skin, which holds the liquid molecules together and tries to minimize its surface area. This property is the surface tension (for a liquid surrounded by gas). The surface tension is responsible for many well-known properties of liquids, *e.g.* the bulge of liquid (the meniscus) above a cup that has been overfilled and the fact that flat stones can be bounced off the surface of a lake. Just as the surface of a liquid has a surface tension, the interface between two immiscible liquids, such as oil and water, has an interfacial tension. This arises because water molecules prefer to be surrounded by other water molecules rather than oil molecules.

If oil and water are vigorously mixed together the oil can be dispersed as an emulsion of small droplets. Small droplets have a large surface area to volume ratio. Consider a test tube containing an emulsion of oil droplets (radius r, total oil volume V_{oil}) in water (Figure 2.1).

The number of droplets (n) is given by the total volume of oil divided by the volume of an individual drop.

$$n = \frac{V_{oil}}{V_{droplet}} = \frac{V_{oil}}{\frac{4\pi}{3}r^3} \tag{2.1}$$

The total interfacial area (A_i) is obtained by multiplying n by the surface area of a droplet.

$$A_i = n \times 4\pi r^2 = \frac{V_{oil}}{\frac{4\pi}{3}r^3} \times 4\pi r^2 = \frac{3V_{oil}}{r} \tag{2.2}$$

Figure 2.1 *Schematic diagram of an oil (light droplets) in water (dark continuous phase) emulsion*

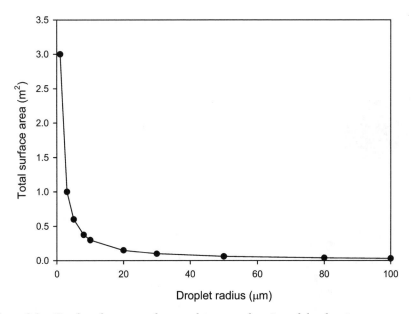

Figure 2.2 *Total surface area of an emulsion as a function of droplet size*

Thus, as the droplets get smaller, the oil–water interfacial area gets larger. This is plotted in Figure 2.2, for $V_{oil} = 1$ cm^3.

Very small droplets have a very large oil–water interfacial area. This means that many oil–water contacts are created, and the interfacial energy, E_i, is large,

$$E_i = \gamma A_i \qquad\qquad (2.3)$$

where γ is the interfacial tension. Emulsions are inherently unstable because they can reduce their energy by reducing the interfacial area, *e.g.* by coalescence of small droplets into large ones. Thus, after some time, a vinaigrette dressing will separate into an oil layer and an aqueous layer. However, emulsions can be stabilized by surface active molecules. These consist of a hydrophilic (water-loving) head and a hydrophobic (water-hating) tail (Figure 2.3a and b). The hydrophilic part of the molecule is attracted to the water and the hydrophobic part is attracted to the oil. The only way to satisfy both parts of the molecule simultaneously is for it to be located at an oil–water interface (Figure 2.3c). This reduces the interfacial tension and makes the emulsion more stable. Experiment 2 in Chapter 8 demonstrates this.

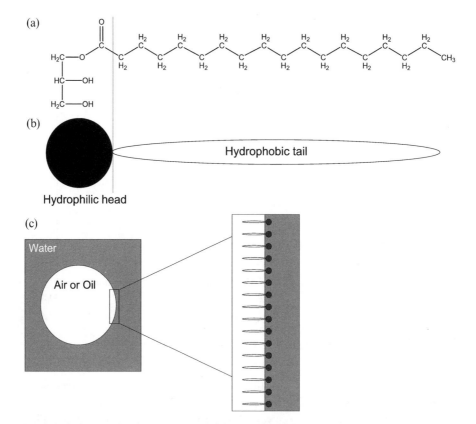

Figure 2.3   (a) *Molecular structure of the emulsifier glycerol monostearate;* (b) *schematic diagram of a surface active molecule; and* (c) *schematic diagram of surface active molecules at the interface between water and air or oil*

Most food emulsions are stabilized by proteins and/or emulsifiers, which slow down or prevent separation of the oil and the water. Proteins are built from amino acids, some of which are hydrophilic and some of which are hydrophobic. Thus certain proteins, such as casein from milk, have hydrophilic and hydrophobic regions, which makes them surface active. Emulsifiers, for example, mono- and diglycerides, also contain hydrophilic and hydrophobic regions.

Sols

Sols are dispersions of solid particles in a continuous liquid phase, *e.g.* the ice crystals in ice cream, pigment particles in ink and paint, soil particles in mud, and platelets in blood. The solid phase is often significantly denser than the liquid phase, so it is liable to sediment out. Brownian motion helps to keep small particles in suspension, and so large particles are most likely to sediment. They can be easily re-suspended by stirring or shaking. This is why you stir a tin of paint before painting, or shake a carton of orange juice before opening it. When the particles are less dense than the liquid, they tend to float to the top of the liquid. This is known as 'creaming', and is more often observed in foams and emulsions, *e.g.* full fat milk where the cream rises to the top of the bottle.

Foams

A foam is a dispersion of gas bubbles in a relatively small volume of a liquid or solid continuous phase. Liquid foams consist of gas bubbles separated by thin liquid films. It is not possible to make a foam from pure water: the bubbles disappear as soon as they are created. However, if surface active molecules, such as soap, emulsifiers or certain proteins, are present they adsorb to the gas–liquid interfaces and stabilize the bubbles. Solid foams, *e.g.* bread, sponge cake or lava, have solid walls between the gas bubbles. Liquid foams have unusual macroscopic properties that arise from the physical chemistry of bubble interfaces and the structure formed by the packing of the gas bubbles. For small, gentle deformations they behave like an elastic solid and, when deformed more, they can flow like a liquid. When the pressure or temperature is changed, their volume changes approximately according to the ideal gas law (PV/T = constant). Thus, foams exhibit features of all three fundamental states of matter. In ice cream, the gas phase volume is relatively low for a foam (about 50%), so the bubbles do not come into contact, and therefore are spherical. Some foams, for example bubble bath, contain so much gas that the bubbles are in close contact, and form a polyhedral structure.

The amount of air incorporated in a foam is often reported in terms of the 'overrun'. The overrun is the ratio of the volume of gas (V_{gas}) to the volume of liquid (V_{liquid}), expressed as a percentage, *i.e.*

$$\text{overrun} = \frac{V_{gas}}{V_{liquid}} \times 100 = \frac{V_{foam} - V_{liquid}}{V_{liquid}} \times 100 \qquad (2.4)$$

Thus for example a foam that has twice the volume of the liquid from which it is made has 100% overrun.

There are several different means of making foams. Whipping is normally used in the kitchen, *e.g.* for foaming cream or egg white. During whipping, large bubbles of air are entrained in a viscous liquid, elongated by stresses due to the vigorous agitation, and as a result break down into smaller ones. The viscosity of the liquid is important. If the liquid is too viscous, it is difficult to beat and therefore to incorporate the air; if it is not viscous enough, the film between the air bubbles rapidly drains, and the bubbles coalesce. The overrun achieved typically increases with whipping speed, until a plateau is reached when equilibrium is established between the rate of bubble formation and the rate of break up. Industrial processes often need to be more reproducible and controllable than whipping. For example, bubbles may be formed at an orifice, *e.g.* sparging air into the liquid. (In ice cream manufacture, air is injected as large bubbles, which are reduced in size by beating.) Bubbles can also be formed *in situ*, *e.g.* in carbonated soft drinks, where bubbles of dissolved gas come out of solution when there is a change of pressure; or in bread, where carbon dioxide is generated by yeast during baking.

Liquid foams, like emulsions, have a tendency to separate into distinct gas and liquid phases in order to decrease the total interfacial area. They may exist for a few seconds (*e.g.* champagne bubbles) or months (*e.g.* ice cream, provided it is kept frozen), depending on the properties of the liquid and the surface active molecule. Creaming in foams is accompanied by drainage of the liquid from between the bubbles. Since the matrix is very viscous, creaming and drainage are very slow in ice cream.

Coarsening of Colloidal Dispersions

Physical systems tend towards the state of lowest energy. We have seen in the section on emulsions above that the interfacial area, and hence the interfacial energy, of a dispersion increases as the particles get smaller. Colloidal dispersions (such as the three in ice cream) therefore have an inherent tendency, driven by surface tension, to separate into distinct bulk phases. While the total volume of the dispersed phase is constant, the size of the particles increases and their number reduces, thus decreasing the total interfacial energy. This is known generally as coarsening, or as recrystallization in the context of ice crystals.

There are two main coarsening mechanisms common to different types of dispersion (though they are known by different names). The first is called coalescence in the context of emulsions and foams, and accretion in the context of dispersions of ice crystals. This is the joining together of two or more adjacent particles to form a single, larger one (Figure 2.4a). The second mechanism is called Ostwald ripening when referring to ice crystals and emulsions, and disproportionation when referring to foams. This takes place by the transfer of individual molecules from small particles to larger ones by diffusion through the continuous phase (Figure 2.4b). Coalescence/accretion is the dominant process for some dispersions, often at high dispersed phase volume, whereas Ostwald ripening/disproportionation is more important in others, such as low dispersed phase volume.

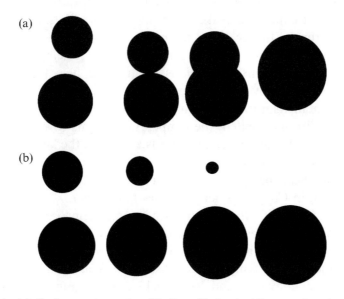

Figure 2.4 (a) *Coalescence/accretion.* (b) *Ostwald ripening/disproportionation*

The recrystallization of ice crystals and the coarsening of air bubbles both lead to deterioration in the texture of ice cream, and a number of methods are employed to slow them down. In contrast, the coalescence of fat droplets is deliberately promoted during ice cream manufacture, and is crucial in making good quality ice cream. The means used to control the coarsening processes of the ice, fat and air dispersions in ice cream are discussed in Chapters 4 and 7.

FREEZING

Ice is a crystalline solid, in which the molecules are held in an hexagonal lattice by intermolecular forces. This hexagonal symmetry explains why snowflakes have 6 sides or points. The molecules vibrate, but stay fixed their positions in the lattice (Figure 2.5a). If the temperature is raised, these vibrations become larger until, at the melting point (0 °C for pure water at atmospheric pressure), the molecules have enough energy to escape from their fixed positions, *i.e.* the ice melts. In the liquid state, the molecules can move past each other, although they remain in close contact (Figure 2.5b).

Supercooling and Nucleation

Figure 2.6a shows the how the temperature of a beaker of crushed ice, initially at −6 °C, changes as it warms up in a laboratory at room temperature. Heat enters the beaker from the surroundings, slowly raising the temperature of the ice. After about 10 min, the temperature reaches 0 °C and the ice begins to melt. At this point, the temperature stops rising and remains constant. This is because the heat that enters the beaker is used up in overcoming the intermolecular forces in the ice lattice as the ice melts, rather than in raising the temperature. Eventually, when all the ice has melted, the temperature begins to rise again. The heat needed to change the ice to water is known as the latent heat. (The heat that causes the temperature of a substance to change is known as the 'sensible' heat, because it can be sensed as a temperature change.) Experiment 6 in Chapter 8 demonstrates latent heat.

Now observe what happens when we reverse the process. Figure 2.6b shows how the temperature of a beaker of water, initially at 5 °C, changes on cooling to below 0 °C. As heat flows out of the beaker into the surroundings, the temperature falls: the molecules have less thermal energy and so move around less quickly. If freezing were the exact reverse of melting, we would expect that, when the temperature reaches

(a)

(b)

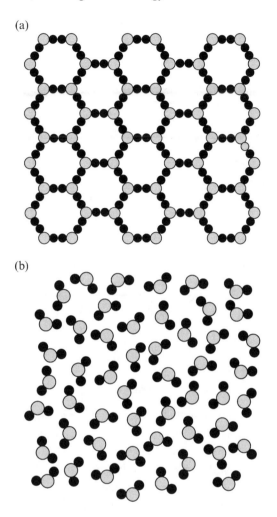

Figure 2.5 *Arrangement of H₂O molecules in* (a) *solid ice and* (b) *liquid water; oxygen atoms are grey and hydrogen are black*

0 °C, ice would begin to form. This would release the latent heat, so that the temperature would then remain constant, until all the water had turned into ice. In fact, the temperature continues to fall below 0 °C, without the formation of ice, in this case to about −2 °C. This phenomenon is called supercooling. At about 20 min, the temperature rises sharply, reaching 0 °C, where it again forms a plateau, before cooling again.

To understand why supercooling occurs, we need to look at what is taking place at a molecular level. When ice crystals melt, the

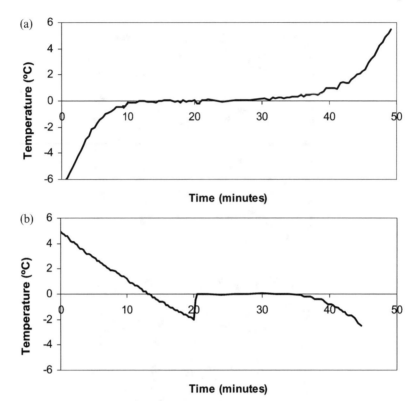

Figure 2.6   *Temperature as a function of time for* (a) *ice melting and* (b) *water freezing*

molecules at the crystal surface switch from the ordered lattice into the liquid one at a time. The reverse process occurs when an ice crystal grows: individual water molecules from the liquid join onto the ice crystal lattice. However, if there are no ice crystals in the beaker, the supercooled water does not have a lattice to join onto and cannot freeze one molecule at a time. So before freezing can occur, tiny embryonic ice crystals ('nuclei') have to form. In the liquid state, nuclei form when the random motion of the water molecules brings a small number of molecules together in a crystal-like arrangement. When a nucleus is formed, an interface is created between the nucleus and the water. There is an energy gain in forming a nucleus because below 0 °C ice has a lower energy than water. However, there is also an energy cost, due to the creation of the new interface. The net energy E of forming a spherical nucleus of radius r is the sum of the energy gain due to change of water to ice and the energy cost due to formation of the interface (equation 2.5).

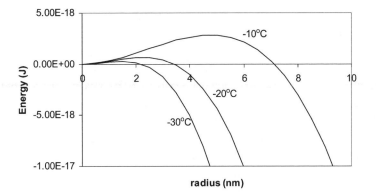

Figure 2.7 *Energy required to form a ice crystal nucleus as a function of ice crystal size at −10, −20 and −30 °C*

$$E = \frac{L(T - T_{\mathrm{m}})}{T} \times \frac{4\pi}{3} r^3 - \gamma \times 4\pi r^2 \tag{2.5}$$

L is the latent heat per unit volume, T is the absolute temperature, T_{m} is the melting point and γ is the surface tension. Figure 2.7 plots this equation for three different temperatures. Below the maximum in the curves, the energy cost of increasing in size is greater than the energy gain. Therefore, small nuclei rapidly break up after they have formed. The maximum in the curves determines the critical radius (r^*) above which nuclei can lower their energy by growing. As the temperature decreases from −10 to −30 °C (*i.e.* the supercooling increases) the critical radius decreases from 5 to 1.5 nm. The height of the maximum, *i.e.* the energy barrier to nucleation also decreases. Thus, the greater the supercooling, the more likely it is that ice crystals will form.

Since nucleation relies on the random clustering of water molecules it is a random process and does not occur at a fixed temperature. Absolutely pure water can theoretically be supercooled to −40 °C. In practice, nucleation occurs at higher temperatures (−2 °C in Figure 2.6b). This is because tiny particles in the water (*e.g.* dust) or the walls of the beaker can act as templates on which the water molecules can begin to cluster in a crystalline arrangement. This means that fewer water molecules are needed to form a stable nucleus, and therefore the supercooling is reduced. Some particles are particularly effective at causing nucleation, for example silver iodide crystals, or a protein produced by the bacterium *Pseudomonas syringae* that grows naturally on plants. In fact, these are often added to the water used in snowmaking machines in ski resorts, to ensure that the water droplets that are sprayed into the cold air freeze before they hit the ground. Experiment 5 in Chapter 8 demonstrates supercooling and nucleation.

As we have seen, energy (latent heat) is needed to melt the ice crystal lattice. Similarly, the reverse process, freezing, gives out latent heat. The latent heat of water is very large because of the strong forces between the water molecules: the heat released when 1 kg of ice freezes is equivalent to that needed to raise the temperature of 1 kg of water from 0 to 80 °C. The release of the latent heat after nucleation causes the rapid rise in temperature that is seen in Figure 2.6b. Once the temperature reaches 0 °C, it remains constant until all the water has frozen. At this point the temperature begins to fall again.

Growth

While nucleation requires several degrees of supercooling, crystal growth (sometimes called propagation) is possible with very little supercooling (<0.01 °C). This is because molecules can join onto an already existing crystal, rather than having to form a completely new one. Crystal growth therefore begins as soon as nucleation has occurred, and continues until equilibrium is reached, *i.e.* the supercooling has been removed. The competition between nucleation and crystal growth determines the characteristics of the ice crystals formed. Rapid freezing (*i.e.* large supercooling/low temperature) produces many nuclei, and allows little time for growth, so numerous small ice crystals are formed. Slow freezing (low supercooling/relatively high temperature), however, produces fewer, larger crystals. Fast and slow freezing processes are used in water ice manufacture to produce different sized ice crystals, and hence different textures. This is discussed in Chapters 4 and 7.

When all the water has frozen, or when there is no longer any supercooling, growth ceases and the amount of ice does not increase any further. However, the system is not necessarily at equilibrium. Small ice crystals have a greater surface area (for a given volume) than large ones, which costs energy. The energy can be lowered if the ice crystals undergo recrystallization.

Freezing Point Depression

The equilibrium freezing point of pure water at atmospheric pressure is 0 °C. When a solute, (*e.g.* sugar or salt) is present, the solute molecules do not fit comfortably into the ice crystal lattice. They effectively get in the way of the water molecules trying to join onto the crystal, so that it

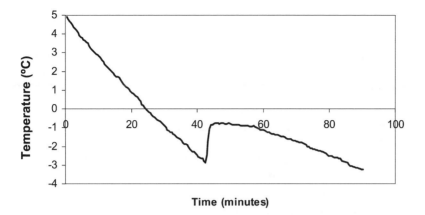

Figure 2.8 *Temperature of a 12% sucrose solution as a function of time during freezing*

is harder for the water to freeze. This results in a lower freezing point. For example, salt is put on roads in winter because it lowers the freezing point to below the temperature of the road, so that the ice melts.

Figure 2.8 shows the same cooling experiment as Figure 2.6, but using a 12% sucrose solution instead of pure water. Supercooling takes place until nucleation occurs, and the temperature rises to the freezing point (in this case −0.77 °C). Unlike pure water, the plateau does not remain flat for long. The formation of ice leads to an increase in the concentration of the sucrose solution, and hence a further decrease in the freezing point, known as freeze concentration.

The amount by which the freezing point changes is known as the 'freezing point depression', and depends on the number of solute molecules present, but not their type. It can be shown that (for low solute concentrations) the freezing point depression, ΔT, is given by

$$\Delta T = Kx \tag{2.6}$$

where K is a constant (known as the cryoscopic constant) and x is the molality of the solution, i.e. the number of moles of solute per kilogram of water. Since the freezing point depression depends on the number of molecules, the smaller the molecular weight of the solute, the more effective it will be at depressing the freezing point on a weight for weight basis. Thus, 1 g of salt (molecular weight 58.5) causes a larger freezing point depression than 1 g of sucrose (molecular weight 342). In fact, determining the freezing point depression caused by a known mass of a compound was historically an important method for measuring its molecular weight. The antifreeze used in cars usually contains ethylene glycol to depress the freezing point. Alcoholic drinks have freezing points below 0 °C, because alcohol (ethanol) lowers the freezing point. This is why vodka can (and indeed should) be served straight from the freezer. Experiment 4 in Chapter 8 demonstrates freezing point depression.

Salt–Water Phase Diagram

We can plot the freezing point of a solution as a function of its concentration. (Throughout this book concentrations are expressed as % weight/weight, abbreviated w/w, unless otherwise stated). This is shown for salt (sodium chloride) solutions in Figure 2.9 (*i.e.* the line separating the saltwater and ice + saltwater regions). The solubility curve of salt in water, *i.e.* the maximum concentration that can be dissolved as a function of temperature, is also shown (the line separating the saltwater and NaCl crystals + saltwater regions). These two curves, together with the horizontal line through their intersection, form the phase diagram for salt solutions. Mixtures of salt and water can take different forms, or 'phases', *i.e.* ice, salt crystals and salt solution. The phase

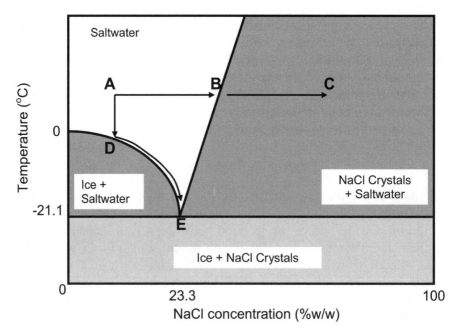

Figure 2.9 *Phase diagram for sodium chloride solutions*
(Reprinted with permission from IOP Publishing Ltd.[1])

diagram is essentially a map that shows which phase(s) are present at any given temperature and composition. However, they do not indicate how the phase is distributed, *e.g.* few large ice crystals or many small ones.

At warm temperatures and low salt concentrations all the salt crystals dissolve in the water: for example, a 15% salt solution at 10 °C (the point marked A). If salt is added while the temperature is kept constant (arrow ABC) the salt concentration increases until, when the solubility line is reached (B), no more salt can be dissolved; the solution is said to be saturated. If any further salt is added it does not dissolve, but remains as crystals, so that the system is a mixture of saturated salt solution and salt crystals (C). Now consider what happens if we start at A and reduce the temperature without adding or removing salt (arrow AD). As the solution is cooled, nothing happens until the freezing point curve is reached at D. If the temperature is lowered below D, ice forms after supercooling and nucleation. The salt is excluded from the ice crystal so the concentration of salt in the solution increases by freeze-concentration. The freezing point is depressed further and the system moves down the freezing point curve, forming more ice. When the freezing point curve meets the solubility curve, the eutectic point is reached (E). This corresponds to 76.7% water or ice, 23.3% sodium chloride and −21.1 °C. The solution cannot be concentrated any further because the limit of solubility of the salt has been reached. If a solution with the eutectic composition is cooled, nothing happens until the whole solution freezes at the eutectic temperature. A solid with the eutectic composition melts at the lowest temperature of any composition; hence the name eutectic, from the Greek for 'easily melted'.

When ice and salt at 0 °C are mixed, some of the ice melts because the salt depresses the freezing point. To melt, the ice must absorb the latent heat from the surroundings. When salt is placed on an icy road, the latent heat is absorbed from the road. Since the road is large, the heat it gives up only causes its temperature to drop by a tiny amount. However, if this is done for example in an insulated bowl the heat can only come from the ice–salt mixture itself. The removal of this heat causes a drop in the temperature. As the latent heat of melting of ice is large compared with the amount of heat required to cause a change in temperature of 1 °C, this causes a significant drop in the temperature. The eutectic temperature can be reached by mixing salt and ice in the correct proportions. This effect was the chief method of refrigeration for producing ice cream in the 19th century, and can be used to make ice cream at home or in a classroom (see Experiment 9 in Chapter 8).

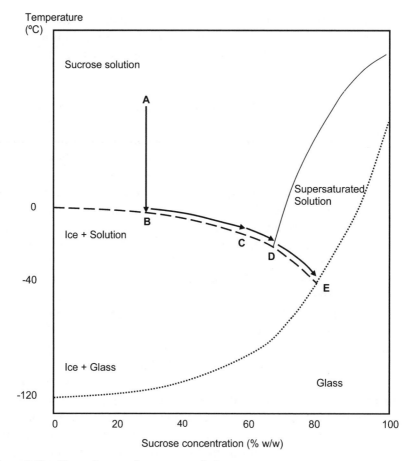

Figure 2.10 *Phase diagram for sucrose solutions*

Sucrose–Water Phase Diagram

The phase diagram for sucrose and water (Figure 2.10) is rather different from the salt–water one, and is particularly important in ice cream making. When a 30% sucrose solution (point A) is cooled, the freezing point curve is reached at B (the freezing point depression deviates significantly from equation 2.6 for sucrose concentrations above about 10%). Ice then forms, the solution freeze concentrates and the freezing point is depressed further (B–C).

At any temperature below the freezing point, (*e.g.* − 10 °C) there is a certain amount of ice in equilibrium with concentrated sucrose solution. The amount of ice can be determined from the phase diagram. At C, the sucrose concentration is 57%. Let us suppose that we started with 1 kg of solution, *i.e.* 300 g sucrose and 700 g water. The sucrose concentration is given by the mass of sucrose divided by the mass of solution, *i.e.*

$$\frac{300}{1000 - m_{ice}} \tag{2.7}$$

where m_{ice} is the mass of ice present at $-10\,°C$ in grammes. We know from the phase diagram that the sucrose concentration is 57%. Therefore

$$m_{ice} = 1000 - \frac{300}{0.57} = 470 \; g \tag{2.8}$$

i.e. a 30% sucrose solution contains 47% ice at $-10\,°C$.

In fact, we can do this calculation to obtain the amount of ice at any temperature, provided we know the freezing point curve. This is important when formulating ice cream recipes, to ensure that the ice cream contains the right amount of ice at its serving temperature. This is often called the ice curve. Figure 2.11 shows the ice curve for a 30% sucrose solution between 0 and $-25\,°C$. The temperature at which the first ice begins to form ($-2.7\,°C$) is the freezing point depression.

So far, the behaviour is the same as with salt solutions. However, unlike salt, sucrose crystals do not form readily, and the solution can become supersaturated, *i.e.* the solute concentration increases beyond the point at which it should precipitate out of solution (*cf.* supercooling) and the freezing point curve can be extended beyond the theoretical eutectic point, D (63% sucrose, $-13.7\,°C$). The solution passes D and continues along the freezing point curve until it meets the

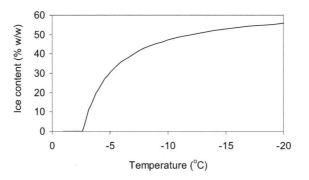

Figure 2.11 *Ice content of a 30% sucrose solution as a function of temperature*

glass transition line (E). As the temperature is reduced and the solution becomes more concentrated, its viscosity increases. Eventually the viscosity becomes so large that the solution effectively becomes a solid. However, unlike a crystal, the molecules are not ordered on a lattice; but have a liquid-like structure, although they are not free to move past each other. The solution forms a solid state known as a glass (ordinary glass has this type of disordered structure). The change to a glassy solid is known as the glass transition. Unlike freezing, it is not a true phase transition because it is a kinetic, not a thermodynamic phenomenon. Nonetheless, a curve representing the glass transition temperature as a function of concentration can be added to the phase diagram. (Technically, it should now be called a state diagram, because it includes a non-equilibrium state.) No more changes happen when the solution is cooled further. In practice it is difficult to reach the maximum freeze-concentration at E because the molecular motion becomes very slow before this point is reached.

Newton's Law of Cooling

Newton's law of cooling states that the temperature difference between a refrigerant (T_r) and (for example) an ice cream mix (T_{mix}) determines the rate at which it cools down (equation 2.9).

$$\frac{dT_{mix}}{dt} \propto (T_r - T_{mix}) \tag{2.9}$$

Thus colder refrigerants cool faster. The coldest refrigerant available to the Victorians was ice and salt at about $-20\,°C$. Today ice cream factories typically use liquid ammonia at $-30\,°C$, and ice cream making is much faster. In fact, Newton's law of cooling explains why the world record for the fastest ice cream ever made used liquid nitrogen at $-196\,°C$.[2] This is demonstrated in Experiment 12 in Chapter 8.

RHEOLOGY OF SOLUTIONS AND SUSPENSIONS

The study of the flow properties of liquids is called rheology. The thickness or runniness of liquids is characterized by their viscosity. Consider a liquid between two parallel surfaces (with area A). The bottom surface is fixed and the top one moves at a constant velocity (v). A force (F) is applied to the top plate to keep it moving. The liquid is dragged along with the moving plate due to its viscosity with a velocity that is largest close to the top plate and decreases with the distance from it. This

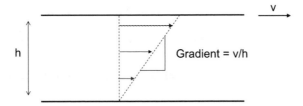

Figure 2.12 *Flow of a Newtonian liquid between parallel surfaces*

kind of deformation is known as 'shear'. Thus there is a velocity gradient in the liquid (v/h), where h is the distance between the plates (Figure 2.12).

Newton suggested that the velocity gradient is proportional to the shear stress σ (*i.e.* the force applied to keep the plate moving per unit area).

$$\sigma = \frac{F}{A} = \eta\frac{v}{h} = \eta\dot{\gamma} \qquad (2.10)$$

The constant of proportionality in equation 2.10 is the viscosity of the liquid (η). Some fluids, such as water, olive oil and sucrose solutions obey this equation and are said to be Newtonian. Their viscosity does not depend on the velocity gradient, *i.e.* how fast the liquid is sheared – known as the shear rate, $\dot{\gamma}$. More complex fluids (*e.g.* solutions of polymers) have a viscosity that does depend on the shear rate. Such fluids are called 'non-Newtonian'. Many complex fluids, for example tomato ketchup and ice cream mix, become less viscous when they are sheared and are described as 'shear-thinning'. Tapping the bottom of the bottle applies shear to the ketchup, which becomes less viscous and flows more easily onto your plate. Other fluids, such as a concentrated solution of cornstarch or quicksand, become more viscous (*i.e.* they are 'shear-thickening'). Experiment 7 in Chapter 8 gives some examples of non-Newtonian fluids. A single viscosity is not sufficient to describe the flow properties of non-Newtonian liquids and if a viscosity is stated, the shear rate at which it was measured must also be given.

The rheology of ice cream is much more complex than that of a simple liquid. The matrix is a solution of small (sugar) and large (stabilizer) molecules, in which particles of other phases (ice crystals, fat droplets and air bubbles) are suspended. We must first look at the effects of each of these, and temperature, in order to understand the rheology of ice cream.

Rheology of Solutions of Small Molecules

The molecules in a liquid are in close contact (Figure 2.5b). The ease with which they can move past each other depends on the forces between them. The lower the temperature of the liquid, the slower the molecules move, and the harder it is for them to overcome these forces and move past each other. The result of this at the macroscopic level is that the whole liquid flows less easily, *i.e.* its viscosity increases. Conversely, as the temperature increases the molecules move faster and more freely so the viscosity decreases. This is illustrated in Figure 2.13, which shows the viscosity of a 20% sugar solution as a function of temperature.

Figure 2.14 shows that the viscosity of sucrose solutions increases as a function of solute concentration.

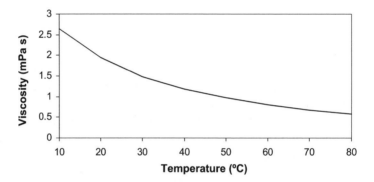

Figure 2.13   *Viscosity of a 20% sucrose solution as a function of temperature*
(Data from ref. 3)

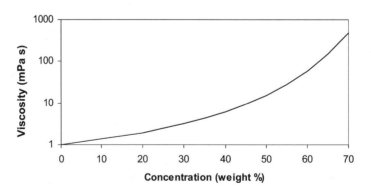

Figure 2.14   *Viscosity versus sucrose concentration at 20 °C*
(Data from ref. 3)

One way to understand this is by considering the sucrose molecules (which are significantly larger than water molecules) as particles in suspension in the water. Einstein showed that the viscosity of a dilute suspension of non-interacting spherical particles (η) in a liquid of viscosity (η_i) increases with the fraction of the volume they occupy (ϕ).

$$\eta = \eta_i(1 + 2.5\phi) \tag{2.11}$$

Einstein applied equation 2.11 to sucrose solutions and, by combining this with data on diffusion, was able to deduce the size of the sucrose molecule.[4] For high concentrations (such as those typical of ice cream mixes) the interactions between the particles further increase the viscosity; this can be accounted for by adding terms proportional to ϕ^2, ϕ^3... to the right-hand side of equation 2.11. Experiment 7 in Chapter 8 describes a simple method for comparing the viscosity of solutions of different concentrations.

Rheology of Polymer Solutions

A polymer is a molecule that consists of a long chain of small molecules (monomers) that are linked together. The stabilizers used in cream are mostly polysaccharides, *i.e.* polymers of sugar molecules. Stabilizers have a number of functions, one of which is to increase the viscosity of ice cream mixes. Figure 2.15 shows the viscosity of a typical polymer solution as a function of concentration. The polymer increases the solution viscosity at low concentrations, and, above a certain concentration, the viscosity increases even more rapidly.

Figure 2.16 shows why this happens. At very low concentrations, each polymer chain takes the form of a separate coil and can move without interference from the others (Figure 2.16a). The coil occupies a larger volume than the total volume of the monomers; this is why the viscosity is higher than for small molecule solutes. However, above a certain concentration (known as the entanglement concentration), the polymer chains begin to overlap and become entangled (Figure 2.16b). This means that it is much harder for the chains to move past each other, and the viscosity increases rapidly. A bowl of spaghetti demonstrates this effect: it can be difficult to pick up one piece without also taking several others because they are entangled. The onset of entanglements can be seen in Figure 2.15 as the increase in the slope at the entanglement concentration. The stabilizers in the matrix of ice cream are usually above the entanglement concentration. The method of Experiment 7 in Chapter 8 can also be used to measure the viscosity of polymer solutions.

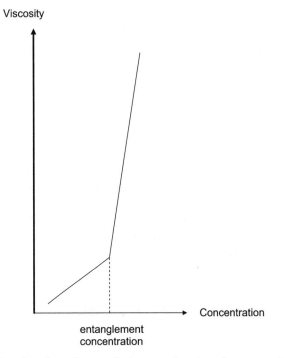

Figure 2.15   *Viscosity of a polymer solution as a function of concentration*

If very high shear is applied to an entangled polymer solution the polymers can be pulled apart and disentangled. The results in a decrease in the viscosity at high shear rates, known as shear-thinning. Figure 2.17 shows shear-thinning in a solution of guar gum, a natural polymer commonly used as a stabilizer in ice cream.

Entanglements are purely topological. However, there can also be specific chemical interactions between polymer chains. For example, there is hydrogen bonding between polymers in guar gum solutions, which results in a higher viscosity than would be expected for purely topological entanglements. Very strong interactions can result in the formation of a cross-linked polymer network, *i.e.* a gel (Figure 2.18). Because the cross-links are fixed, the polymers cannot move past each other. Therefore, instead of flowing like a liquid, the gel returns to its original shape when the shear is removed, *i.e.* it is elastic. (The analogy in this case is spaghetti that has been knotted together, so that the strands cannot be separated.) Several stabilizers can form gels under certain conditions, *e.g.* locust bean gum, pectin, sodium alginate and κ-carrageenan.

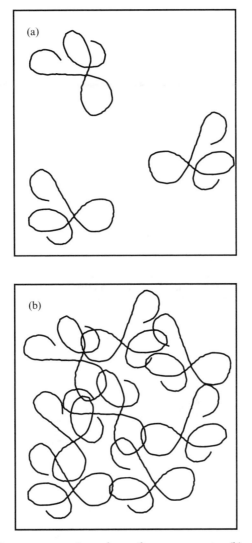

Figure 2.16 (a) *Low concentration: the coils are separate;* (b) *above the critical concentration, they overlap and entangle*

Rheology of Suspensions

Ice crystals increase the viscosity of ice cream because they are solid particles suspended in the matrix (*cf.* equation 2.11). However, because the ice crystals can interact during flow (since their volume fraction is quite high) and because they are not spherical, the viscosity is greater than equation 2.11 predicts. Similarly, fat droplets in emulsions and gas bubbles in foams with a relatively low gas-phase volume can also be

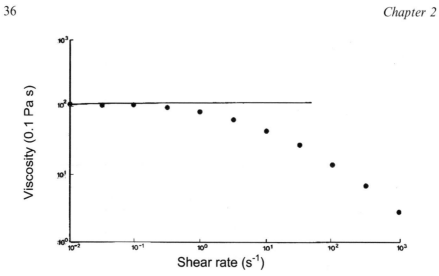

Figure 2.17   *Viscosity (η) of a guar solution as a function of shear rate (γ̇)*
(Reprinted from 'Viscosity – Molecular Weight Relationships, Intrinsic
Chain Flexibility, and Dynamic Solution Properties of Guar Galactoman-
nan',[5] Copyright, 1982, with permission from Elsevier)

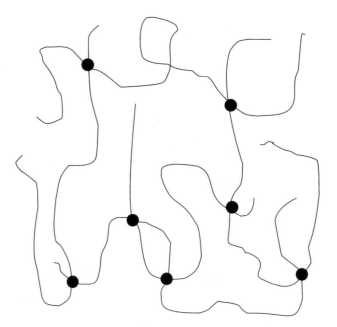

Figure 2.18   *Polymer gel, with permanent cross-links between the polymer strands*

considered as suspended particles. Foam rheology becomes more complex at high gas-phase volumes because bubbles interact with each other. Air bubbles and liquid fat droplets are not rigid and therefore can deform in the flow. At very high shear, they can be broken up. This is important at two points in the ice cream manufacturing process: the break up of the fat emulsion in the homogenizer and the air bubbles in the factory freezer (Chapter 4).

The combination of the effects of temperature, small and large molecules in solution, and hard and soft suspended particles (whose number and size change through the manufacturing process) makes the rheology of ice cream extremely complex. This is an area which has been, and continues to be, the subject of much research. The rheology of ice cream and the techniques used to measure it are described in Chapters 6 and 7.

REFERENCES

1. C.J. Clarke, *Phys. Educ.*, 2003, **38**, 248.
2. 'The Guinness World Records 2002', Guinness World Records, London, 2002.
3. M. Mathlouthi and J. Génotelle, in 'Sucrose: Properties and Applications' ed. M. Mathlouthi and P. Reiser, Blackie Academic & Professional, London, 1995.
4. A. Einstein, *Ann. Phys.*, 1906, **19**, 289 and 1911, **34**, 591.
5. G. Robinson, S.B. Ross-Murphy and E.R. Morris, *Carbohydr. Res.*, 1982, **107**, 17.

FURTHER READING

E. Dickinson, 'An Introduction to Food Colloids', Oxford University Press, Oxford, 1992.

J.W. Mullin, 'Crystallization', 3rd Edition, Butterworth-Heinemann, Oxford, 1997.

Chapter 3

Ice Cream Ingredients

The ingredients of ice cream products can be classified in three groups:

- Major ingredients, present in substantial quantities (at least a few % by weight), such as milk protein, sugar, fat and water.
- Minor ingredients, present in small quantities (less than about 1% by weight), such as emulsifiers, stabilizers, colours and flavours.
- Components such as chocolate, biscuits, wafers, fruit pieces and nuts that are combined with ice cream to make products.

Most ice creams also contain a significant proportion (by volume) of air, although this is not usually thought of as an ingredient. The ingredients can be obtained from various raw materials: for example, milk protein and fat (and some water) could be supplied together in the form of milk or cream; alternatively, they could come from separate raw materials, *i.e.* skimmed milk powder and butterfat or vegetable fat. This choice largely depends upon the type of product required, cost and availability of raw materials and the scale of production. Table 3.1 shows a typical ice cream composition (or 'formulation').

The total solids is the sum of all the ingredients other than water. In general, high total solids formulations give high quality ice cream. In an 'all-natural' ice cream, artificial emulsifiers, colours and flavours are avoided. Products, such as sorbets, milk ices or water ices contain a subset of the ingredients of ice cream. For example, water ices do not usually contain milk protein or fat. A typical water ice formulation is given in Table 3.2. Concentrated fruit juice (added at a few percent) may be used instead of the colours, flavours, acid and some of the sugar.

Table 3.3 gives a nutritional analysis of a 100 ml serving of a typical ice cream. Ice cream is a good source of essential amino acids from the

Table 3.1 *Typical ice cream formulation*

Ingredient	Amount (wt%)
Fat	7–15
Milk protein	4–5
Lactose	5–7
Other sugars	12–16
Stabilizers, emulsifiers and flavours	0.5
Total solids	28–40
Water	60–72

Table 3.2 *Typical water ice formulation*

Ingredient	Amount (wt%)
Sugars	14–24
Stabilizers, colours and flavours	0.5
Citric acid	0.5
Total solids	15–25
Water	75–85

Table 3.3 *Nutritional analysis of a typical ice cream*

Ingredient	Grammes per 100 ml ice cream
Total fat	7
Saturated fat	5
Carbohydrate	14
Sugars	13.5
Protein	1.8
Fibre	0.5
Energy	530 kJ (125 kcal)

milk proteins, vitamins and minerals. Its fat and sugar content make it a high energy density food. Numerous ice cream products with nutritional benefits have been developed in recent years. These include reduced fat or reduced sugar ice cream, cholesterol-free ice cream, ice cream enriched with vitamins, calcium or fibre, ice cream containing polyunsaturated fats, and prebiotic or probiotic ice cream that promotes the 'good' bacteria in the intestinal tract.

MILK PROTEINS

Cows' milk contains about 87% water. The remainder consists of fat (4%), proteins (3.5%), lactose (4.8%) and small quantities of inorganic salts, notably calcium and phosphate (0.29%). The exact composition depends on the breed of cow, the diet and the season. Fat and lactose are discussed below in the sections on fats and oils, and sugars, respectively. The components of milk other than fat and water are collectively known as milk solids non-fat (MSNF), since they are supplied together in whole milk or skimmed milk powder.

Milk contains two main types of protein: casein (80%) and whey proteins (20%). Casein and whey proteins are distinguished by their solubility at pH 4.6 (at 20 °C): caseins are insoluble, whereas whey proteins are soluble. There are four main casein proteins: α_{s1}, α_{s2}, β and κ casein. Most of the casein proteins are present as colloidal particles, typically 100 nm in size, known as casein micelles. Their natural function is to carry the insoluble calcium phosphate needed by mammalian young. Casein micelles scatter light; this accounts for the opacity of skimmed milk. The caseins are relatively small proteins. They are very surface active, because one end of the molecules consists mostly of hydrophilic amino acids (such as serine and glutamic acid) whereas the other consists mostly of hydrophobic ones (for example, leucine, valine and phenylalanine). Caseins are quite stable to heat denaturation, but can be denatured by excessive heat, leading to aggregation and precipitation. There are also four types of whey protein: lactoglobulin, lactalbumin albumen and immunoglobulins. These are globular proteins that are also surface active. They are more heat sensitive than the caseins, and lose their surface activity on heat denaturation. Milk also contains one further type of protein, enzymes. These are a group of proteins that can catalyse specific chemical reactions.

Milk proteins have two important functions in ice cream. Firstly, they can stabilize water-continuous emulsions and foams because they are surface active. We will see in Chapter 4 that this has important consequences for the formation and stability of the air bubbles in ice cream. Secondly, they contribute to the characteristic dairy flavour. Milk proteins for ice cream manufacture are obtained from several different raw materials:

- milk (concentrated, skimmed or whole)
- skimmed milk powder
- whey powders
- buttermilk or buttermilk powder.

To produce concentrated or powdered milk, whole milk is first pasteurized and then separated into skimmed milk and cream that contain 0.1% and 48–50% fat respectively. The skimmed milk is then concentrated to a defined composition by evaporation. This can be spray dried to produce skimmed milk powder.

Whey is a cheap source of milk protein since it is a by-product of cheese manufacture. It is often used in a powdered form. However, it has some disadvantages in ice cream manufacture. Firstly, it can increase the amount of lactose in the formulation. At high lactose concentrations, the lactose can come out of solution and form crystals that produce a sandy texture in ice cream (Chapter 7). For this reason, it is preferable to use whey powders in which the lactose content has been reduced. Secondly, as mentioned above, whey is less heat stable than casein and can be denatured during the manufacturing process, reducing its functionality.

Buttermilk is a by-product of butter manufacture, in which pasteurized cream is cultured and then churned to produce butter and buttermilk. It has approximately the same composition as skimmed milk, and can also be concentrated and spray dried. Buttermilk provides a distinctive, fresh flavour.

The choice of source of milk protein is based on availability, convenience and cost. Liquid products offer ease and speed of transfer and weighing, whereas powders do not need chilled storage, have a more consistent composition and lower transport costs.

SUGARS

Sugars are used in all types of ice cream and water ice. A whole range of different molecules, such as glucose, fructose (the sugar in fruit), sucrose (the sugar you have in your kitchen) and lactose (the sugar in milk) are covered by the term 'sugar'. Monosaccharides are the simplest group of sugars, and conform to the chemical formula $(CH_2O)_n$. The most important group of monosaccharides are the hexoses, *i.e.* those for which $n = 6$. There are many different molecules with this formula (isomers), of which the naturally occurring ones are dextrose, fructose, galactose and mannose. When two monosaccharides are joined together, a disaccharide is formed. Naturally occurring disaccharides include sucrose, the major soluble energy reserve in plants, trehalose, which has a similar function in fungi, yeasts, lichens and insects, and lactose, which, as already mentioned, is present in mammalian milk. Higher oligosaccharides (*e.g.* raffinose) are formed from three or more monosaccharides. When more than about ten monosaccharides are joined together, the resulting polymer is known as a polysaccharide. Most stabilizers (discussed below) are polysaccharides. The chemistry of saccharides is complex, because as well as the naturally occurring molecules there are many other isomers. Some of these are chrial, *i.e.*

two molecules have the same structure, except that they are mirror images of each other (known as stereoisomers). Only one of the two stereoisomers of glucose occurs naturally, namely dextrose. The large structural diversity of sugars gives them a range of different chemical, physical and sensory properties. Here we focus only on those sugars commonly used in ice cream.

Sugars have two major functions in ice cream. They make it sweet, and they control the amount of ice and hence the softness of ice cream (the higher the ice content, the harder the ice cream). We saw in Chapter 2 that sugars lower the freezing point of solutions and therefore reduce the amount of ice. For example, a standard ice cream has an ice content of about 55% by weight at −18 °C. By changing the amounts and types of sugars, an ice content of 45% can be obtained, giving a softer ice cream, *e.g.* for scooping, with the same sweetness. Sugars can also influence the texture of ice cream in another way because they affect the viscosity of the matrix. The higher the molecular weight of the sugar, the higher is the matrix viscosity. This can have both beneficial and detrimental effects on the ice cream: high viscosity matrices tend to give creamier, warmer eating products but they are also more difficult to scoop.

Dextrose

Dextrose ($C_6H_{12}O_6$) is the simplest sugar (Figure 3.1), and is the naturally occurring stereoisomer of glucose. Dextrose is manufactured by the hydrolysis of starch. It is generally supplied as dextrose monohydrate, which contains approximately 91% dextrose and 9% water. Dextrose is somewhat less sweet than sucrose.

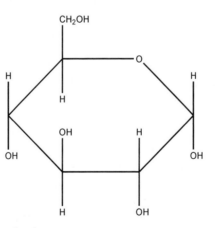

Figure 3.1 *Dextrose molecule*

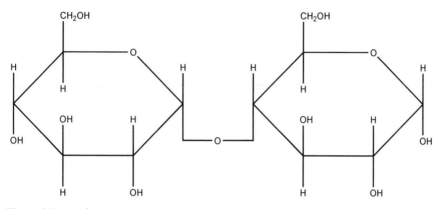

Figure 3.2 *Maltose molecule*

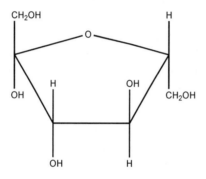

Figure 3.3 *Fructose molecule*

When two molecules of dextrose are linked together they form the disaccharide maltose (Figure 3.2).

Fructose

Fructose ($C_6H_{12}O_6$, Figure 3.3) occurs naturally in fruits and honey. Pure fructose is expensive and little used in commercial food products except for specialized products, *e.g.* for diabetics. Fructose is substantially sweeter than sucrose.

Sucrose

Sucrose ($C_{12}H_{22}O_{11}$) is the sugar most commonly used in ice cream. It is a disaccharide and consists of a dextrose molecule linked to a fructose molecule (Figure 3.4). It is extracted from sugar cane, which grows in

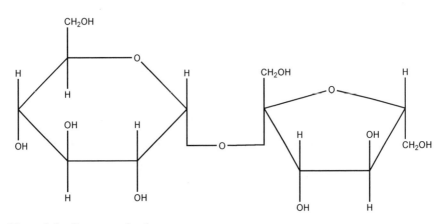

Figure 3.4 *Sucrose molecule*

tropical climates, or sugar beet, which grows in temperate climates. After refining, both sources give the same pure crystalline material. Sucrose can be hydrolysed (*i.e.* the monosaccharides are separated) either by heating with acid or by using the enzyme invertase. The resulting product, an equal mixture of dextrose and fructose, is called invert sugar, which is occasionally used as an ice cream ingredient.

Lactose

Lactose ($C_{12}H_{22}O_{11}$) is a disaccharide of dextrose and galactose and is present in milk (Figure 3.5). Lactose is substantially less sweet than sucrose. It also has a relatively low solubility, as a result of which it can crystallize out of ice cream as a monohydrate (*i.e.* for each lactose molecule there is also a water molecule in the crystal). The crystals

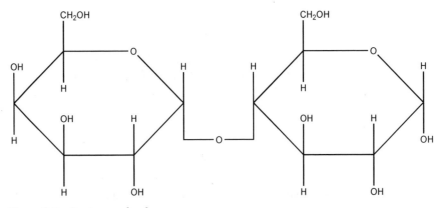

Figure 3.5 *Lactose molecule*

produce a sandy texture, making the product unacceptable (Chapter 7). Lactose can be hydrolysed by the enzyme lactase to give dextrose and galactose. Lactose intolerance, which occurs frequently in certain ethnic groups, is caused by missing or defective lactase. However, many people who are technically lactose intolerant can still consume milk products.

Corn Syrup

Corn syrup (sometimes also called glucose syrup) is a mixture of dextrose, maltose, maltotriose and higher oligomers of dextrose, obtained by the partial hydrolysis of starch with acid and/or enzymes. The exact composition depends upon the hydrolysis conditions. Corn syrups are specified by their dextrose equivalent (DE). The higher the DE, the lower the average molecular weight. Dextrose has a DE of 100 and starch has a DE of 0. Commercial corn syrups have DE in the range 35–65. However, the DE alone does not completely describe the sugar blend as different mixtures of dextrose and its oligomers can have the same DE. The sweetness of corn syrups increases as the DE decreases; most are less sweet than sucrose. Corn syrups are generally supplied as liquids containing about 20% water.

Sugar Alcohols

Sugar alcohols are formed when sugars are reacted with hydrogen in the presence of a catalyst, *e.g.* sorbitol from dextrose, lactitol from lactose, and mannitol from mannose. They are somewhat less sweet than sucrose. They are incompletely digested by the human digestive system and therefore have about half the calorific value of sugars. A negative consequence of this is that they can have a laxative effect when consumed in large amounts. Therefore, their use is largely restricted to specialized products, *e.g.* for diabetics.

OILS AND FATS

The distinction between fats and oils is that the former are solid at room temperature, and the latter are liquid. Ice cream typically has a fat content of 8–10% by weight, though in premium ice creams it can be as high as 15–20%. Fat performs several functions in ice cream: it helps to stabilize the foam, it is largely responsible for the creamy texture, it slows down the rate at which ice cream melts and it is necessary to deliver flavour molecules that are soluble in fat but not water. The major sources of fat used in industrial ice cream production are butterfat, cream and vegetable fat.

The Chemistry of Oils and Fats

Fats are largely made up of triglycerides (98%), together with small amounts of phospholipids and diglycerides. Triglycerides are esters of glycerol (propan-1,2,3-triol) and fatty acids (monocarboxylic acids). The fatty acids determine the physical properties of the triglyceride, such as the melting and crystallization behaviour and viscosity. Fats have the same general chemical structure (Figure 3.6). R_1, R_2 and R_3 can be the same or different fatty acid residues, for example stearic ($R = C_{17}H_{35}$), oleic ($C_{17}H_{33}$) or palmitic ($C_{15}H_{31}$).

There are many naturally occurring fatty acids. They are classified as saturated, which have no carbon–carbon double bonds ($C=C$) in the carbon chain; mono-unsaturated fatty acids, which have one; and poly-unsaturated fatty acids, which contain two or more. Since the molecule cannot twist about a $C=C$ bond there are two different geometric isomers, cis and trans (Figure 3.7). The cis isomer is more common in nature.

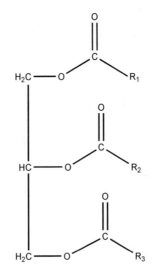

Figure 3.6   *Strucure of triglycerides*

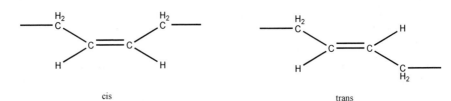

Figure 3.7   *Cis and trans isomers*

Table 3.4 *Approximate fatty acid composition (wt%) of butterfat, palm oil and coconut oil* (Data from ref. 1)

Chain length: C=C bonds	Name	Butter fat	Palm oil	Coconut oil
4:0	Butyric	4		
6:0	Caproic	2		1
8:0	Caprylic	1		5
10:0	Capric	2		8
12:0	Lauric	3		45
14:0	Myristic	11	1	18
16:0	Palmitic	29	40	11
16:1	Palmitoleic	5		
17:0	Margaric	1		
18:0	Stearic	9	6	2
18:1	Oleic	27	43	8
18:2	Linoleic	4	10	
18:3	Linolenic	0		
20:0	Arachidic	2		

A useful convention is to denote fatty acids by the number of carbon atoms and the number of C=C bonds. For example, lauric acid, which has 12 carbon atoms and no C=C bonds, is C12:0. This nomenclature does not specify the position of the C=C bonds, nor whether they are cis or trans. All fats are mixtures of triglycerides (and hence contain a number of different fatty acid residues). The approximate fatty acid composition of some fats is shown in Table 3.4. Butterfat contains a much wider range of fatty acids than the vegetable fats. Coconut oil contains very high levels of saturated fatty acids, particularly lauric acid.

Since fats are mixtures of triglycerides, they do not have a single melting point but actually melt over a range of temperatures. The melting profile, which gives the amounts of solid and liquid fat as a function of temperature, is usually measured by NMR. The fatty acid chain lengths and the degrees of unsaturation influence the melting profile: fats with short chains and high degrees of unsaturation melt at low temperatures. The melting temperatures of unsaturated fats can be raised by hydrogenating the fat, *i.e.* adding hydrogen to convert C=C double bonds into C–C single bonds. This is known as hardening. Fats can also be blended to modify the melting profile.

Good quality ice cream can only be made with fats that have a suitable melting profile. Fats that melt at high temperatures produce ice cream with a waxy mouth-feel. Conversely, it is difficult to create stable foams with fats that melt at low temperatures, for reasons that are discussed in Chapters 4 and 7. Dairy fat has the right melting profile to give ice cream a smooth creamy texture, and also provides dairy flavour. Ice cream can also be made with palm oil and coconut oil because they have fairly similar melting profiles to dairy fat. In some countries ice cream products made with non-dairy fat may not be legally described as ice cream. Nonetheless, the global market for ice cream made with vegetable fat is substantial.

WATER

Water forms a high proportion of ice cream (typically 60–72% w/w) and water ice mixes (typically 75–85%). Water is the medium in which all of the ingredients are either dissolved or dispersed. During freezing and hardening the majority of the water is converted into ice.

EMULSIFIERS

In addition to the milk proteins, ice cream also contains other surface active molecules, namely emulsifiers, such as mono- and diglycerides or lecithin (from egg yolk). Despite their name, emulsifiers, as we will see in Chapter 4, are actually used in ice cream to de-emulsify some of the fat.

Mono-/diglycerides

The most commonly used emulsifiers in ice cream manufacture are mono-/diglycerides (E471). Mono-/diglycerides are mixtures of mono-glycerides and diglycerides. Whereas fats are triglycerides (*i.e.* esters of glycerol with three fatty acid molecules), monoglycerides are esters of glycerol with one fatty acid molecule, and diglycerides are esters of glycerol with two fatty acid molecules (Figure 3.8). Mono- and digly-cerides are surface active because the glycerol end of the molecule is hydrophilic and the fatty acid end is hydrophobic. Just as for triglycer-ides, the fatty acids determine the properties of the mono-/diglycerides.

Mono-/diglycerides are made by partially hydrolysing vegetable fats, such as soybean oil and palm oil. (Animal fat-based emulsifiers are not commonly used because they are not suitable for vegetarian and certain religious diets.) They normally contain 40–60% monoglyceride, together with diglyceride, and a small amount of triglyceride. Fully saturated mono-/diglycerides that contain predominantly stearic and palmitic acids, such as glycerol monostearate, are often used for ice cream production and typically make up about 0.3% of the ice cream mix. Materials with high monoglyceride content (>90%) are available ('himonos'). These are difficult to disperse because they can become

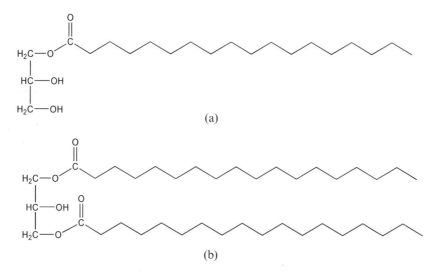

Figure 3.8 *Structure of* (a) *glycerol monostearate (a monoglyceride) and* (b) *glycerol distearate (a diglyceride)*

extremely viscous and form a gel in aqueous systems. However, this may be helpful in some applications. For example, the gelling properties have been exploited in the manufacture of very low fat ice cream. Sorbitan esters of fatty acids, such as polyoxyethylene sorbitan monooleate (also known as polysorbate 80) are structurally similar to monoglycerides. These consist of a fatty acid attached to a sorbitol molecule instead of glycerol. Polyoxyethylene groups are also attached to the sorbitol molecule to make it water soluble. Polysorbate 80 can be used as an ice cream emulsifier, typically at concentrations of 0.02–0.04%.

Egg Yolk

Egg yolk, which contains several components with emulsifying properties, notably lecithin, is often used in 'all-natural', premium or home-made ice creams. Egg yolk has the approximate composition (by weight) of 50% water, 16% protein, 9% lecithin, 23% other fat, 0.3% carbohydrate and 1.7% minerals. Lecithin consists of phosphatides and phospholipids. Egg yolk is usually supplied for use in ice cream manufacture either as pasteurized fresh egg yolk, frozen sugared pasteurized egg yolk (which has had about 10% sucrose added to protect it from damage during freezing) or as dehydrated egg yolk. Egg yolk solids are normally used at about 0.5–3%. High concentrations are only used for super-premium products, and can give the ice cream an eggy flavour, which is seen as beneficial in some markets.

STABILIZERS

Stabilizers are a group of water-soluble or water-dispersible biopolymers used in small amounts (typically 0.2%) in ice cream, sorbets, water ices and other foods. Most stabilizers are polysaccharides of plant origin, *e.g.* alginates and carrageenans (from seaweeds), locust bean gum and guar gum (from tree seeds), pectin (from fruit) and sodium carboxymethyl cellulose (from cotton). Xanthan, a bacterial polysaccharide, and gelatin, a polypeptide of animal origin, are also sometimes used. These biopolymers are polydisperse and polymolecular, because their structures vary with the source and the environmental conditions. Nutritionally, stabilizers are a source of soluble fibre. Although they come from natural sources, under European law they are considered food additives and therefore they have associated 'E numbers'.

Stabilizers are straight or branched polymers containing hydroxyl groups that can form hydrogen bonds to water molecules. Typically they contain ~10^3 monomer units and have molecular weights of ~10^5–10^6. Because they are large, stabilizers do not dissolve in water as readily as smaller molecules: some require high temperatures or shear for complete hydration. When dissolved, they produce high viscosity solutions at low concentrations (Chapter 2). Some stabilizers in solution can form gels when heated and/or cooled or on the addition of cations. Others have complex solution properties, such as shear-thinning behaviour or particularly high viscosities.

Stabilizers have several beneficial effects in ice cream during manufacture, storage and eating. Some of these are non-specific effects, *i.e.* they are achieved by increasing the viscosity of the matrix phase, independent of the type of stabilizer used. Stabilizers can:

- Produce smoothness in texture during eating.
- Reduce the rate of meltdown (*i.e.* the rate at which the ice cream loses mass as it melts).
- Prevent shrinkage and slow down moisture migration out of ice cream during storage.
- Mask the detection of ice crystals in the mouth during eating.
- Allow easier pumping and more accurate filling during processing.
- Facilitate the controlled incorporation of air in the factory freezer and help produce a stable foam.

There are also specific effects due to the particular properties of certain stabilizers. For example, some stabilizers can retard ice crystal growth during storage (Chapter 7). Finally, there are synergistic effects, which arise from the combination of two stabilizers, for example, greater increases in viscosity than would be expected from either component individually, or gelation, when the separate components do not gel.

Sodium Alginate

Sodium alginate (E401) is a polysaccharide of guluronic acid and mannuronic acid, which is extracted from brown seaweeds such as *Macrocystis pyrifera* (found off the Pacific coast of North America, Australia and New Zealand) and *Laminaria digitata* (from Ireland, Norway, France and Scotland). It consists of a negatively charged polymer chain with ionic bonds to positively charged sodium ions (Na^+). In aqueous solution, the sodium ions dissociate from the polymer so it becomes charged (this type of polymer is called a polyelectrolyte). Calcium ions (Ca^{2+}) or other doubly charged cations can bind to negative charges on two different polymer molecules. These intermolecular interactions lead to the formation of a gel. In ice cream, alginates are blended with phosphate, citrate or tartrate ions to prevent premature gelation due to the calcium from the milk solids. The major advantage of alginate is its resistance to acid conditions, particularly when heated, whereas other stabilizers would lose their functionality.

Carrageenan

Carrageenans (E407) are complex polysaccharides of esters of galactose and α-3,6-anhydrogalactose, found in red seaweeds (*Rhodophycae*), such as *Chondrus crispus* (Irish Moss), *Kappaphycus alverezii* and *Eucheuma denticulatum*. Carrageenan can have several structures, usually classified as one of three types that have different properties: *kappa* (κ), *iota* (ι) (both of which come from Indonesia and the Philippines) and *lambda* (λ) (from the Pacific coast of North America, France, Denmark, Norway, Ireland and Great Britain). Carrageenans are polyelectrolytes in aqueous solution and form gels after heating and cooling in the presence of cations such as K^+ or Ca^{2+}. Carrageenans can also form gels with both milk proteins and locust bean gum. κ-Carrageenan has a specific function in ice cream: added at about 0.02%, it reduces phase separation of milk proteins and polysaccharides, a phenomenon known as wheying off (Chapter 7).

Locust Bean Gum

Locust bean gum (E410), also known as LBG, carob gum or St Johns Bread, is extracted from the seeds of the Mediterranean *Ceratonia siliqua* tree. These large, evenly sized seeds were the original carats used as a measure of weight. LBG is a polysaccharide consisting of a mannose backbone with galactose side branches on about a quarter of the mannose units. The side branches occur in blocks, giving LBG 'smooth' regions of free mannose backbone and 'hairy' regions of galactose side groups (Figure 3.9a). LBG is only partially soluble in cold water and it must be heated to $>85\,°C$ to hydrate it fully. In solution, strong hydrogen bonds can form between the large smooth backbone regions (Figure 3.9b). This leads to gel formation under certain conditions. LBG is the best stabilizer for many ice cream applications and its ability to gel is crucial to some aspects of its use (Chapter 7). However, it is also expensive, and subject to large fluctuations in availability and price.

Guar Gum

Guar gum (E412) is extracted from the seeds of *Cyamposis tetragonolobus*, an annual crop grown in the Indian subcontinent. Guar has a similar structure to LBG: it has a backbone of mannose units, about half of which have galactose side branches. Guar has a higher molecular weight than LBG and the side groups are more evenly spaced (Figure 3.9c). The larger proportion of galactose units makes guar cold-water soluble. The regions of backbone that are free of side chains are smaller than in LBG. Hydrogen bonding between them is therefore not strong enough to form permanent cross-links, but does result in 'hyper-entanglements'. These are stronger than purely topological

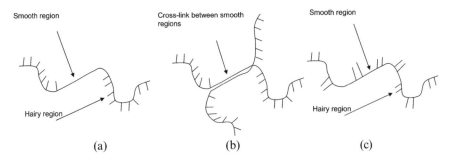

(a)   (b)   (c)

Figure 3.9 (a) *Structure of locust bean gum (LBG);* (b) *a cross-link in an LBG gel formed by hydrogen bonding between smooth regions on the two chains;* (c) *structure of guar gum*

entanglements and account for the high viscosity of guar gum solutions at low concentrations. Guar gum is significantly cheaper than LBG.

Pectin

Pectin (E440) is extracted from citrus peel and apple pomace. It is a polysaccharide consisting of linear chains of galacturonic acid and galacturonic acid methyl ester units. Pectin is classified according to its degree of esterification. High methoxy (>50% esterified) and low methoxy (<50%) pectins possess different properties. For example, low methoxy pectin requires calcium to gel whereas high methoxy pectin gels at low pH and in the presence of high concentrations of sugar. Pectin is the setting agent used in jam making. All fruits contain some pectin. Some fruits, such as apples and gooseberries, usually contain enough natural pectin to form a gel, whereas pectin must usually be added to set the jam for other fruits, such as strawberries and cherries.

Xanthan

Xanthan (E415) is produced by the bacterium *Xanthomonas campestris*. It is a polysaccharide consisting of a chain of glucose residues with charged trisaccharide side groups. Xanthan has excellent solubility and is suitable for use under acid conditions, *e.g.* in water ice. Xanthan is a rod-like polymer. In solution, the rods are oriented in different directions and interact to form a weak network. When a small amount of shear is applied, the rods all line up and the network is broken. When the shear is removed, the network reforms. As a result, the viscosity of xanthan solutions decreases dramatically with shear, but quickly recovers once the shear is removed. This property is useful in sauces for ice cream. During dispensing, the viscosity is low, but as soon as shear forces cease, the viscosity rises substantially. This results in a sauce that stays put after dosing on to the product. However, xanthan is not widely used in ice cream because it is expensive.

Sodium Carboxymethyl Cellulose

Sodium carboxymethyl cellulose (E466) is derived from purified cellulose from cotton and wood pulp. It is a sodium salt polymer of anhydroglucose residues. For use in ice cream, an average of 0.7 of the 3 hydroxyl groups in each glucose unit are substituted with a sodium carboxymethyl group. The long, negatively charged molecules produce a stable thickener that can also reduce casein precipitation. However, its perception as a 'chemical' has resulted in fairly low usage.

Gelatin

Gelatin is a mixture of high-molecular-weight polypeptides derived from collagen from animal connective tissues, and was commonly used as a gelling and thickening agent. It is not suitable for vegetarians, and has now generally been replaced by other stabilizers.

FLAVOURS

An essential requirement of ice cream products is that they taste appealing. The flavours used in ice cream manufacture are usually supplied as solutions of aroma and taste compounds. Some flavour molecules are fat soluble, whereas others are water soluble. This affects the perception of flavour in ice cream: water-soluble flavours are present in the matrix and are released rapidly on consumption, whereas fat-soluble flavours are released more slowly. Flavours may be natural, *i.e.* extracted from sources such as plants, or synthetic. The latter can be nature identical (artificially produced but identical to the naturally occurring form) or artificial (artificially produced and not occurring in nature). They are used to impart flavour to products, to enhance inherent flavours and to ensure uniformity of flavour between batches. Fruit acids, such as citric or malic acid are added to fruit flavoured water ice products to give them extra 'bite', by making them sour. The three most important ice cream flavours are vanilla, chocolate and strawberry.

Vanilla

Vanilla is the most popular ice cream flavour in the UK. It comes from the fruit pods of a large, climbing tropical vine that is a member of the orchid family, which grows in places such as Madagascar, Bali, Java and the French Pacific Islands. Over 50 species of vanilla orchid exist but just three are used commercially: *Vanilla planifolia*, *V. tahitensis* and *V. pompa*. Vanilla pods are picked before they fully ripen. Vanillin (Figure 3.10), the major flavour component of vanilla, is formed during curing, which takes several months. After curing, the pods are oily, smooth, aromatic and very dark brown and contain about 2% vanillin by weight. The pods can either be used whole, ground, or, most commonly, to produce vanilla extract.

Natural vanilla flavours have a much more aromatic and delicate character than pure vanillin because more than 250 components contribute to the flavour. However, because of high demand, vanillin is manufactured by biosynthesis, using micro-organisms to imitate the formation of vanillin during the curing process. Vanillin can also be

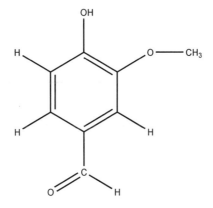

Figure 3.10 *Vanillin molecule*

synthesized from a coal tar derivative or from lignin, a by-product of the paper industry.

Chocolate

Chocolate is the second most popular ice cream flavour in the UK and derives its colour and flavour from cocoa products. These come from the beans of the cocoa tree, *Theobroma cacao*, which grows in countries such as the Ivory Coast, Ghana, Nigeria, Brazil, Ecuador, Malaysia and Indonesia. The Greek word *theobroma* means 'food of the gods'. Pods grow from fertilized flowers out of the main trunk of the cocoa tree and on the thicker branches. When ripe, the pods turn a yellowish purple colour and are harvested. A group of compounds called flavonoids are responsible for the colour. The pods are cut open and the beans and some of their surrounding pulp is removed. The pods contain between 20 and 40 beans, each weighing approximately 1 g. The beans are then left to ferment for several days, during which time the characteristic brown of cocoa is formed as the flavonoids turn into tannins. Short fermentation times produce a light, mild flavoured cocoa, whereas longer fermentation times produce a darker, bitter flavoured cocoa. The complex cocoa flavour has at least 460 components.

The fermented beans are dried under carefully controlled conditions, and are then bagged and transported for processing to produce cocoa liquor. This is a dark brown substance which is liquid above 36 °C. It contains approximately 55% fat and 45% non-fat cocoa solids. Cocoa butter is the fat obtained by pressing hot cocoa liquor. The residue, known as cocoa press cake, is ground up to produce cocoa powder (usually containing 10–12% fat and 88–90% non-fat cocoa solids). The non-fat cocoa solids mainly provide the colour and flavour and the cocoa butter provides the texture and mouth-feel of chocolate.

Chocolate stimulates the release of endorphins in the brain, which provide a feeling of pleasure; it contains caffeine and also phenylethylamine, a stimulant that increases the heart rate, raises the blood pressure and can give rise to feelings of lust or passion.

Strawberry

Strawberry is the third most popular ice cream flavour in the UK. Fresh or fresh-frozen strawberries may be used, although the fruit alone does not generally give a sufficiently intense flavour. Strawberry flavours are available as extracts from the prepared fruit, as artificial compounds or as artificially fortified extracts. To obtain a good strawberry flavour it is necessary to have the correct sugar/acid balance. This can be achieved by adding fruit acids, *e.g.* malic acid or citric acid.

COLOURS

The colour of ice cream has a significant influence on the consumer's perception of its flavour and quality. Colours are added to ice cream for several reasons:

- To give colour to products that would otherwise be virtually colourless (*e.g.* water ice products).
- To reinforce colours already present, *e.g.* from fruit.
- To ensure uniformity of colour between different batches.

Natural colours extracted from plants have been used as colouring agents for foodstuffs for many years. Synthetic colours based on petro-chemical products were developed in the 20th century. Natural colours have a healthy image and good solubility, but may be expensive (partly because they need to be used in high concentrations) and can suffer from poor stability to heat and light. Artificial colours, such as azo dyes, attract adverse publicity in some countries, whereas in others they are considered acceptable. (Both natural and artificial colours have E numbers.) Commonly used natural colours include anthocyanins (E163) which are red–purple and come from black grapes, elderberries, red cabbages and hibiscus; chlorophylls and chlorophyllins (E140), which are green–yellow and come from green leaf plants, such as nettle and spinach; turmeric (E100), which is yellow; and vegetable carbon black

(E153), which is black and comes from carbonized vegetable material. Cocoa powder is also used as a colouring agent in ice cream.

OTHER COMPONENTS

A number of other components, such as chocolate, fruit, nuts and bakery products are used to add value and interest to ice cream or to make products such as choc ices and ice cream cones.

Chocolate

There are three main types of eating chocolate, all of which consist of small solid particles dispersed in cocoa butter. It is important that the particles are the correct size (10–25μm): if they are too small, the chocolate is slimy; if they are large, the chocolate is gritty. Table 3.5 shows a typical plain (dark) chocolate composition for use in ice cream (this is not the same as a normal dark chocolate because it is eaten at a lower temperature – see Chapter 5).

In milk chocolate, about 10% skimmed milk powder replaces some of the sugar and cocoa powder. White chocolate contains sugar and milk solids but no cocoa powder. All three types are used as coatings, toppings *etc.* in ice cream products. The composition of chocolate is legally defined in most countries: it must contain cocoa butter, and the amounts and types of other fats and emulsifiers that it can contain are restricted. Chocolate analogues that fall outside these requirements are known as couverture. Couverture has similar eating properties to chocolate but, since other fats such as coconut oil can be used, couvertures have a wider range of textures and may be cheaper. To make a couverture that melts in the mouth when eaten, the fats should be chosen so that they melt below 30 °C: any higher and a waxy mouth-feel results. Couvertures can also be flavoured, for example, with lemon, strawberry or yoghurt. As well as providing flavour, colour, and

Table 3.5 *Composition of a typical plain chocolate for use in ice cream*

Ingredient	Amount (wt%)
Cocoa butter	50–55
Cocoa powder (non-fat)	10–15
Sugar	30–35
Lecithin	0.5

texture contrast, chocolate and couverture are widely used in ice cream products, in four main ways:

- External coatings, for example for choc-ices, where it gives a clean and easy to hold surface, and chocolate-coated stick products.
- Inclusions, for example pieces of chocolate added to tubs of ice cream.
- Toppings and decorations, for example sprinkled onto ice cream cones.
- Moisture barriers: a chocolate coating is sprayed onto the inside of the cone in products like *Cornetto*, to prevent water from the ice cream migrating into the wafer, thereby preserving its crispness.

These are discussed in Chapter 5.

Fruit

Fruit is used in ice cream products, either as fruit pieces, or as a sauce. Fruit pieces add novelty to ice cream products and enhance the perception of healthiness. Sauces, often fruit, but also chocolate and toffee, are widely used as toppings on cone and cup products or as ripples in ice cream. They provide flavour and texture contrast as well as an attractive product appearance. A wide range of fruits is used in ice cream products, from common European fruits such as strawberry and apple, to tropical fruits such as mango and banana.

Most fruit preparations contain 60–70% fruit pieces, and fruit sauces generally contain 25–40% fruit. Wherever possible, the fruit is quick-frozen near to the time and place of harvest. Some fruits are diced or chopped before use, whilst others are used whole. Typically, fruit pieces are steeped in a sugar solution for several hours or days at elevated temperatures ($>80\,^{\circ}\text{C}$) with gentle stirring. This infuses sugar into the fruit pieces to ensure that they do not become hard, icy and unpalatable at frozen temperatures. Stabilizers (typically LBG, guar or pectin) are added to give the fruit sauce the high viscosity required for product assembly. The preparation is subsequently cooled to about $4\,^{\circ}\text{C}$ over several days, during which time colours, flavours and flavour enhancers can be added. Citric acid serves as a flavour enhancer for acidic fruit flavours and can also reduce enzymic browning of the fruit during processing and storage. Colourings may be used if the fruit does not have a strong colour that is retained during processing. One difficulty with this preparation process is that it can be very destructive to the fragile fruit pieces, resulting in fruit preparations that have a very jammy flavour and texture, quite unlike fresh fruit.

Nuts

Nut pieces are added to ice cream products to provide texture contrast, flavour and visual appeal. Several types of nuts are used, for example almonds, cashews, hazelnuts, peanuts, pecans, pistachios and walnuts. Walnuts and pecans are often candied to create a crisper texture and darker colour. In the candying process, the nuts are mixed with sugar and then cooked at about 115 °C. Nut pastes can be used as flavourings. Nut ingredients are generally expensive. Since some people are allergic to nuts control systems must be put in place to ensure that nuts cannot enter products on which they are not declared as ingredients.

Bakery Products

Two main types of baked goods are used in ice cream products: wafers and biscuits. Ice cream wafers typically have a composition of 40% wheat flour, 17% sugars, 37% water, 5% oil, 1% lecithin and 0.1% salt. They have a greater sugar content than confectionery wafers. This gives them greater plasticity when hot, so they are more suitable for the rolling and folding processes required to create shapes such as cones. Biscuits are used to make sandwich products, or in small pieces as inclusions in ice cream. A typical composition consists of 67% flour, 20% sugar and 13% fat.

REFERENCES

1. 'CRC Handbook of Chemistry and Physics 76th Edition', ed. D.R. Lide, CRC Press, Boca Raton, CA, 1995.

FURTHER READING

R.T. Marshall, H.D. Goff and R.W. Hartel, 'Ice Cream', 6th Edition, Chapman & Hall, New York, 2003, Chapters 3 and 4.

S.T. Beckett, 'The Science of Chocolate', The Royal Society of Chemistry, Cambridge, 2000.

C. Fisher and T.R. Scott, 'Food Flavours: Biology and Chemistry', The Royal Society of Chemistry, Cambridge, 1997.

Chapter 4

Making Ice Cream in the Factory

Simply mixing together the ingredients and freezing them does not make good quality ice cream, because this does not produce the micro-structure of small ice crystals, air bubbles and fat droplets held together by the matrix. The factory process for making this microstructure is based on the same principles as the Victorians used, but it has evolved over many years to enable the rapid and cost-effective production of ice cream on an industrial scale. The details of the process vary from factory to factory, according to the type of equipment and scale of manufacture, but the basic stages, which are described in this chapter, are common to all. These are: mix preparation (which consists of dosing and mixing of the ingredients, homogenization and pasteurization), ageing, freezing and hardening. The process is summarized in Figure 4.1, which also shows the points of addition of the ingredients, including air, and the temperature changes that take place. The assembly and packaging of products by combining ice cream with other components, such as cones, chocolate, fruit, sticks, tubs *etc.* takes place before and/or during the hardening stage, and is covered in Chapter 5.

MIX PREPARATION

Dosing and Mixing of Ingredients

The first step in the manufacture of ice cream is the preparation of the mix. The mixing process is designed to blend together, disperse and hydrate (dissolve) the ingredients in the minimum time with optimal energy usage. The ingredients must be dosed in accurate amounts in a particular order to achieve optimum and consistent mix quality and maximum utilization of the ingredients. Bulk ingredients are usually dosed into the mix tank automatically, whereas smaller quantities are weighed out and tipped in manually. The mix tank has a means for

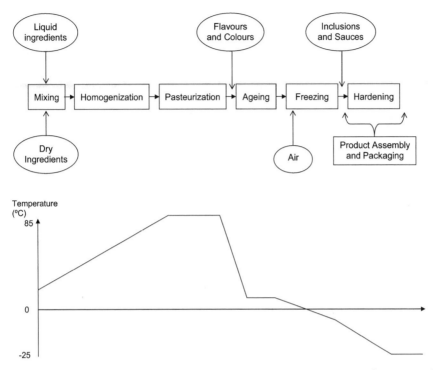

Figure 4.1 *Schematic diagram of the manufacturing process, showing the points of addition of ingredients and components and the temperature profile*

heating the mix, a stirrer to help mixing, and is usually insulated to minimize heat losses. Heating and stirring are carefully controlled, so that the ingredients are effectively dispersed and dissolved, and that heat-sensitive ingredients are not damaged.

The liquid ingredients (water, milk, cream *etc.*) are dosed first, and heating and agitation commence. Solid fats are melted before addition. Dry ingredients (sugars, stabilizers, milk powder *etc.*) are added next. Stabilizers are the most difficult ingredients to dissolve. To aid dissolution they are dry mixed with at least an equal weight of sugar, prior to addition to the mix tank. This mixture is added slowly to the tank to ensure even dispersion and avoid the formation of lumps of incompletely dissolved stabilizer. Milk powder and whey powder are also added slowly to prevent the formation of lumps. The outlet from the mix tank contains a filter to remove any lumps that do form. Milk products, particularly whey proteins, denature at high temperatures so the mix should not be hotter than 85 °C when they are added. Colours and flavours are usually added at this stage, unless they are heat sensitive, in which case they are introduced after pasteurization. For various

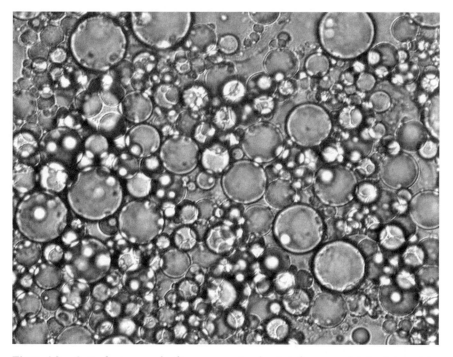

Figure 4.2   *Optical micrograph of ice cream mix, showing the coarse fat droplet emulsion*
produced in the mix tank prior to homogenization
(The image is 225 µm wide)

reasons, such as the start-up and shutdown of factory freezers or interruptions in production, some mix is not converted into saleable ice cream. This can be re-used by adding it back into the mix tank (provided that it is the same formulation as the new mix, or that the formulation is adjusted accordingly). This is known as rework.

Once all of the ingredients have been added, the mix should be homogeneous and at or above about 65 °C. The shear forces from the stirring produce a coarse oil-in-water emulsion with relatively large fat droplets (about 10 µm in diameter), shown in Figure 4.2.

Homogenization and Pasteurization

The mix is next pasteurized to reduce the number of viable microorganisms to a level that is safe for human consumption, and homogenized to break the fat particles down into many small droplets. Figure 4.3 shows a flow diagram of a factory pasteurization and homogenization process. Much energy is needed to heat the mix up to the pasteurization temperature. To maximize the energy efficiency,

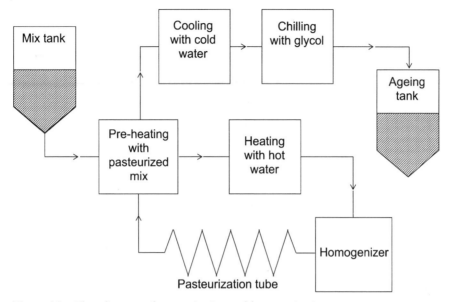

Figure 4.3 *Flow diagram of pasteurization and homogenization*

heating occurs in two stages. In the first stage, the mix is taken from the mix tank and passed through a plate heat exchanger (Figure 4.4). This is designed to ensure good heat transfer and to allow easy cleaning, and consists of two separate flow systems that pass through alternate plates. The first flow system contains the incoming mix, and the second contains hot mix that has already been pasteurized and homogenized. Thus the incoming mix is heated and the pasteurized and homogenized mix undergoes the first stage of cooling, which is necessary before the next step in the manufacturing process.

In the second heating step, the mix is further heated with hot water in another section of the plate heat exchanger. At the end of this stage the mix must be hot enough to ensure that the pasteurization temperature is achieved after homogenization. However, the temperature should not exceed 85 °C in order to prevent denaturation of the milk proteins and to avoid the introduction of off-flavours (for example the taste of cooked milk).

In the homogenizer the hot mix (>70 °C) is forced through a small valve under high pressure (typically up to about 150 atm). The large fat droplets are elongated and broken up into a fine emulsion of much smaller droplets (about 1 µm in diameter), greatly increasing the surface area of the fat (Figure 4.5). Sometimes a second homogenization step is used with a lower pressure (about 35 atm) to reduce clustering of the small fat droplets after the first stage. Recently, very high pressure

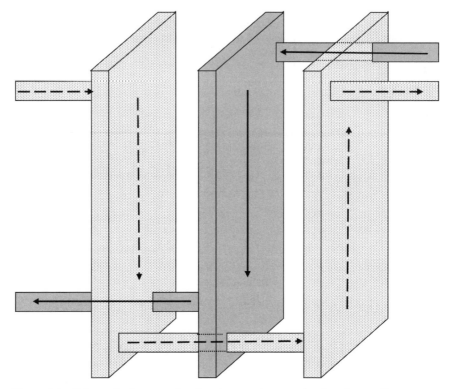

Figure 4.4 *Schematic diagram of a plate heat exchanger, showing the two flow systems*

homogenization (up to 2000 atm) has been shown to produce even smaller fat droplets, and therefore a higher fat surface area for a given volume. This makes more efficient use of the fat, which can have a number of benefits. For example, in low fat mixes this makes the air bubbles more stable, and hence reduces the rate of meltdown. Figure 4.6 shows the effect of homogenization on a standard ice cream mix.

After homogenization, the milk proteins readily adsorb to the bare surface of the fat droplets. The proteins are mostly adsorbed on the aqueous side of fat–matrix interface, with hydrophobic parts at the interface. Free casein, casein micelles and whey have different surface activities, so they adsorb differently onto the fat droplets; for example, casein adsorbs more than whey. Proteins are very good at stabilizing oil-in-water emulsions against coalescence because they provide a strong, thick membrane around the fat droplet. Interactions between the proteins on the outside of the droplets make it harder for the droplets to come into close contact. This is known as steric stabilization.

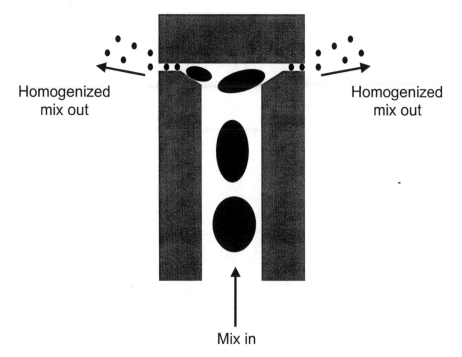

Homogenized
mix out

Homogenized
mix out

Mix in

Figure 4.5 *Homogenization: large fat droplets are forced through a small valve under high pressure, elongating them and breaking them up*

Pasteurization takes place in the holding tube. This is a pipe from the homogenizer outlet, whose length and diameter are chosen to ensure that the mix is held at the pasteurizing temperature for the required time. The temperature of the mix coming out of the holding tube is measured to ensure that all of the mix has been subjected to the minimum time/temperature combination. A typical pasteurization regime is a temperature of 80.5 °C and a holding time of 31 s.

After pasteurization the mix is cooled in three steps. First, heat is transferred to incoming mix from the mixing tank (as described above); then it is cooled with water, which to save energy can be later used for mixing or cleaning; finally, it is cooled to 4 °C with chilled glycol. Cooling inhibits microbial growth and prepares the mix for ageing.

AGEING

The cooled mix is pumped to the ageing tanks. These are designed to minimize the exposure of mix to the atmosphere and to other possible sources of contamination. Mix is held at between 0 and 4 °C and is

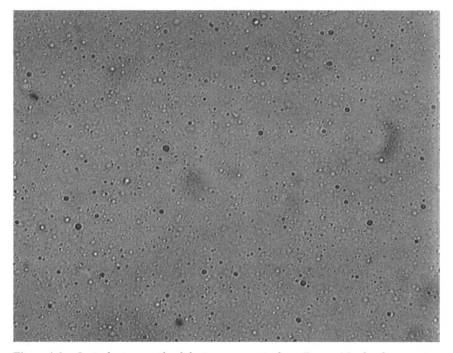

Figure 4.6   *Optical micrograph of the ice cream mix from Figure 4.2 after homogeniza-*
tion, showing the fine fat droplet emulsion
(The image is 225 μm wide)

gently stirred from time to time. The minimum amount of agitation is
applied in order to avoid heating the mix. Heat-sensitive ingredients,
such as colours, flavours and fruit purées, may be added at this
stage. Of course any material added after pasteurization must be
microbiologically safe to avoid contaminating the pasteurized mix.

Two important processes take place during ageing. First, the emulsifiers adsorb to the surface of the fat droplets, replacing some of the milk
protein (Figure 4.7). This is aided by the fact that as the mix cools
the mono-/diglycerides begin to crystallize, which makes them more
hydrophobic, so that they adsorb more strongly onto the fat droplets.
The emulsifiers have their fatty acid chains in the fat phase, and their
polar heads at the surface. Displacement of some of the protein by
emulsifiers produces a weaker membrane. This is strong enough to
stabilize the emulsion under the static conditions in the ageing tank, but
makes the emulsion unstable under shear.

Second, the fat inside the droplets begins to crystallize. Crystallization is slow because nucleation must occur inside each individual
droplet. Crystalline mono-/diglycerides and high melting point

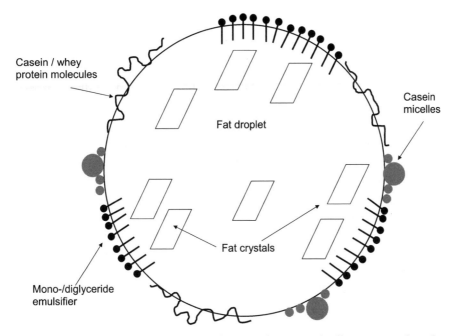

Casein / whey
protein molecules

Casein
micelles

Fat droplet

Fat crystals

Mono-/diglyceride
emulsifier

Figure 4.7 *A fat droplet during ageing, showing adsorption of milk proteins and emulsifiers (not drawn to scale) at the surface of the fat droplet and crystallization of the fat*

triglycerides promote fat crystallization by acting as nucleation points. Fat crystals may protrude through the droplet surface. It is essential that ageing is long enough for crystallization to occur and for emulsifiers to displace some of the protein since both of these processes are important precursors to the next stage in ice cream production. Without them, it is difficult to incorporate and stabilize air bubbles when the mix is frozen in the factory freezer. The ageing time, hence the extent of fat crystallization and emulsifier adsorption, depends on the nature of the mix and on the purpose for which it is to be used. Mixes intended for extruded products must be aged for a minimum of 6 hours because, as we will see below, this leads to greater partial coalescence, and hence stiffer ice cream. Two hours is sufficient for most products. It is often convenient to age a mix overnight, and hold it in the ageing tanks until it is required for production, but this should not be for longer than three days.

During ageing the mix is sampled for laboratory analysis. This typically includes measurements of the mix viscosity and the amount of fat and total solids. Stabilizers, which are slow to dissolve completely in water, continue to hydrate and swell during ageing. The apparent viscosity of the mix is often measured as a quality control test. If the mix does not have the expected viscosity, there may be a problem in the mixing or ageing processes, for example the stabilizers have not fully hydrated. Microbiological safety is checked after ageing. Plate counts are used to test for pathogenic and other potential spoilage micro-organisms, such as *Listeria monocytogenes* and *Salmonella*. A small sample of mix or ice cream is spread on a plate containing a suitable growth medium, such as agar gel, and incubated, for example at 32 °C for 48 hours. The plate is viewed under a microscope. Colonies appear as spots on the plate and are counted, from which the number of bacteria per gram of ice cream is calculated. The count must be below a certain number specified by legislation. Provided the ice cream mix is pasteurized properly and that the rest of the manufacturing process is carried out under hygienic conditions there should be no micro-biological problems in the factory, or indeed up to consumption since micro-organisms can not grow in ice cream that is kept frozen.

FREEZING

So far, we have only formed one part of the microstructure of ice cream, the fat droplets. The other parts are created in the next stage, freezing, which is the core of the manufacturing process. The factory ice-cream freezer converts mix into ice cream by simultaneously aerating, freezing and beating it, to generate the ice crystals, the air bubbles and the matrix. Simultaneous aeration, freezing and beating has been the basis of ice cream production since its invention and remains so today, both in factory ice cream freezers and domestic ice cream makers.

Modern factory ice cream freezers belong to a class of equipment known to chemical engineers as scraped surface heat exchangers; these are designed to remove heat from (or add heat to) viscous liquids. Ice cream freezers consist of a cylindrical barrel typically 0.2 m in diameter and 1 m long. (However, factory freezers designed for different production rates have barrels with a wide range of sizes.) A refrigerant, normally a liquefied volatile gas, such as ammonia or Freon, flows through a jacket (Figures 4.8 and 4.9) and cools the outside of the barrel as it evaporates. Inside the barrel is a rotating stainless steel dasher driven by an electric motor. The dasher is equipped with scraper blades that fit very closely inside the barrel. The dasher has two functions: to subject the mix to high shear and to scrape off the layer of ice crystals that forms on the very cold barrel wall (hence the term 'scraped surface heat exchanger'). The barrel is often made from nickel, covered

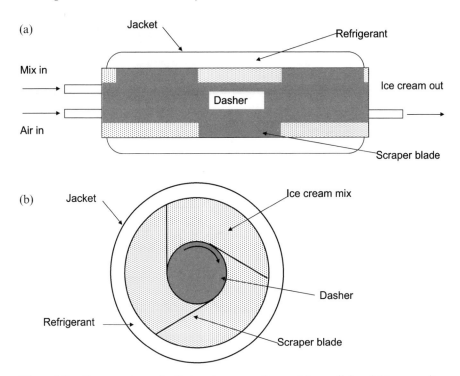

Figure 4.8 *Cross-sections of a factory ice cream freezer* (a) *parallel and* (b) *perpendicular to the axis of the barrel*

on the inside with a thin layer of chromium. Nickel gives good heat transfer, and can withstand high pressures. The chromium coating provides resistance to wear from the scraping, and chemical resistance to the cleaning agents used between batches.

There are two types of dasher, open and closed, which are used for different types of product (Figure 4.10). Open dashers have an open cage supporting the scraper blades, within which is a passively rotating whipper. The dasher occupies 20–30% of the volume of the barrel. Closed dashers have a solid core, and occupy approximately 80% of the volume. Open dashers give lower shear and longer residence times than closed ones for the same outlet temperature and throughput. Open dashers are generally used for ice cream production. The longer residence time helps to achieve good aeration. Closed dashers are used when a low throughput is required (*e.g.* for sauces and ripples) or for ice cream that needs to retain its shape after extrusion, because, as we will see below, the higher shear increases the amount of partially coalesced fat and makes the ice cream stiffer.

(a)

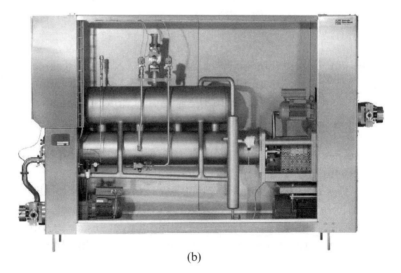

(b)

Figure 4.9   (a) *External view of an industrial ice cream freezer*; (b) *internal view (the lower cylinder is the jacketed barrel)*
(Kindly provided by WCB Ice Cream)

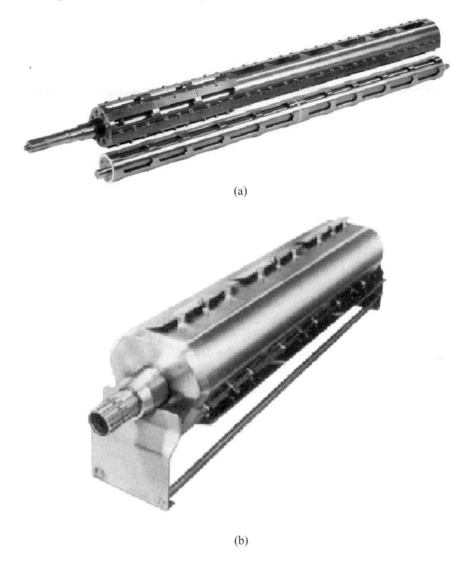

(a)

(b)

Figure 4.10 (a) *An open dasher*; (b) *a closed dasher*

Ice cream mix at approximately 4 °C is pumped from the ageing tank into the barrel, where it is aerated and frozen, before being pumped out from the other end. The operation of the factory freezer is controlled by several parameters. The pressure of the refrigerant sets the temperature at which it evaporates, and hence the wall temperature (typically $-30\,°C$). The mix and air in-flow and ice cream out-flow rates determine the time that the mix spends inside the barrel (known as the residence time, typically 30–60 s), the overrun, the pressure inside the barrel (typically 5 atm) and the throughput (which can be as much as 3000 l h^{-1} in a large industrial freezer). All of these, together with the dasher rotation speed (typically 200 rpm), determine the outlet temperature. Modern factory freezers are computer-controlled, allowing easy monitoring and control of the process parameters.

Air is injected into the barrel through a system of filters to ensure that it is clean, dry and free from microbiological contamination. Initially the air forms large bubbles. It is essential to create (and maintain) a dispersion of small air bubbles to obtain good quality ice cream. The beating of the dasher shears the large air bubbles and breaks them down into many smaller ones: the larger the applied shear stress, the smaller the air bubbles. Long residence times also lead to small air bubbles. It is easier to whip air into a foam that consists of a large volume fraction of liquid and a small volume fraction of air than *vice versa*. The high pressure inside the barrel reduces the volume of the air that has been introduced, and therefore makes it easier to aerate further.

The shear also causes some of the fat droplets to collide and coalesce. The mixed protein–emulsifier layer is designed so that the emulsion is stable under static conditions but unstable under shear. Even though mono-/diglycerides and lecithin are called emulsifiers, their function in ice cream is to *de-emulsify* the fat. Needle-like fat crystals protruding from the droplet surface help this process. They can puncture the mixed protein emulsifier layer on the other droplet, allowing the droplets to coalesce. Figure 4.11 shows three possible results of two fat droplets colliding. If all the fat is liquid the droplets coalesce completely to form a single spherical droplet (Figure 4.11a). However, if all the fat is solid the droplets cannot coalesce at all (Figure 4.11c). The choice of fat type and the ageing process ensure that some of the fat in an ice cream mix is solid so the droplets can *partially* coalesce, *i.e.* they form a cluster

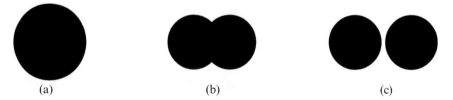

(a)   (b)   (c)

Figure 4.11   *Fat droplet coalescence*: (a) *liquid fat droplets coalesce completely*; (b) *fat droplets containing some solid fat partially coalesce and retain some of their separate identities*; (c) *completely solid fat droplets cannot coalesce*

that retains some of their individual nature (Figure 4.11b). Partially coalesced fat droplets are also known as de-emulsified or destabilized fat.

Partial coalescence helps to stabilize the air bubbles in ice cream. Some of the fat droplets (discrete and partially coalesced) come into contact with the air–matrix interface during freezing. These, together with the surface active milk proteins, provide the air bubbles with some stability against coalescence. The balance between fat, protein and emulsifier are critical for the manufacture of ice cream because it controls the stability of the emulsion and hence the ease of aeration and the stability of the air bubbles. If there is an excess of protein the emulsion may be too stable, so that not enough fat is destabilized. This produces an unstable foam, and the ice cream is coarse and wet. Conversely, if the emulsion is too weak, for example due to too much emulsifier or too little protein, too much fat is de-emulsified and the coalesced fat droplets become so large that they are detectable on eating. This is known as 'buttering'. The amount of destabilized fat is therefore an important parameter both for quality control and research.

In a standard ice cream formulation, sufficient partial coalescence occurs to enable a stable air cell structure to be maintained at overruns up to about 120%. It can be difficult to obtain overruns of more than about 60% in products where fat and protein are not present, or only present in small quantities, such as sorbets. Similarly it is difficult to obtain high overruns if there is insufficient shear (for example because of a very short residence time) to produce partially coalesced fat. Extra shear, and hence increased de-emulsification, can be produced either by increasing the dasher speed or by using a closed dasher.

Whilst the mix is aerated, it is simultaneously frozen. A significant difference between Victorian freezers and modern ones is the refrigerant. The best refrigerant available to the Victorians was a eutectic mixture of ice and salt, and the lowest temperature that could be achieved was -21.1 °C. Modern factory freezers commonly use liquid ammonia at about -30 °C. As it absorbs heat from the ice cream mix, it boils and evaporates. It is then sucked out, reliquefied, cooled and returned to the jacket. According to Newton's law of cooling (equation 2.9), the colder the refrigerant the faster the heat removal from the mix, and hence the faster the rate at which ice cream can be made.

When the mix touches the cold barrel walls, a layer of ice is immediately formed, and is rapidly scraped off by the rotating scraper blades (Figure 4.12). The time between scrapes is typically 0.1 s, so only a very thin layer of ice can form before it is removed. It is important that the scraper blades are kept in good condition to ensure efficient removal of ice. Even a very thin layer of ice remaining on the barrel wall can cause

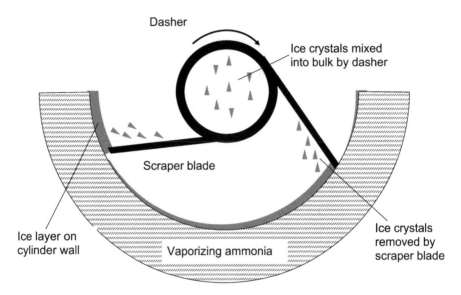

Dasher

Ice crystals mixed
into bulk by dasher

Scraper blade

Ice layer on
cylinder wall

Vaporizing ammonia

Ice crystals
removed by
scraper blade

Figure 4.12 *Scraper blades removing ice crystals from the barrel wall*

a dramatic reduction in heat transfer. The small ice crystals that are scraped off are dispersed into the mix by the beating of the dasher. There are large temperature gradients inside the barrel, both radial (from colder at the wall to warmer at the centre) and axial (from warmer near the inlet and colder towards the outlet). Therefore, when ice crystals are dispersed into the warmer mix at the centre of the barrel, they melt and cool it down. Near the inlet the crystals all melt, but about one-third of the way through the barrel the mix becomes cold enough for the ice crystals to survive.

Heat must be extracted from the mix both to cool it down (the sensible heat) and to freeze water into ice (the latent heat). We can estimate their relative contributions as follows. Consider a 1 kg mass (m) of a 20% sucrose solution as a simple model for ice cream. This enters the factory freezer at $+4.0\,^{\circ}\mathrm{C}$ (T_1), and leaves at $-5.0\,^{\circ}\mathrm{C}$ (T_2). The specific heat capacity (c), *i.e.* the amount of heat required to change the temperature of 1 kg by 1 °C, of a 20% sucrose solution is 3.5 kJ $\mathrm{kg^{-1}\,K^{-1}}$. (Although this varies with temperature it is very nearly constant between $+4$ and $-5\,^{\circ}\mathrm{C}$.) The sensible heat removed (ΔQ_S) is therefore

$$\Delta Q_S = mc(T_1 - T_2) = 1.0 \times 3.5 \times 10^3 \times (4.0 - -5.0) = 3.2 \times 10^4 \ \mathrm{J} \qquad (4.1)$$

The latent heat of fusion of ice (L) is 330 kJ $\mathrm{kg^{-1}}$. A 20% sucrose solution has a freezing point of $-1.5\,^{\circ}\mathrm{C}$ and at $-5\,^{\circ}\mathrm{C}$ the ice content is

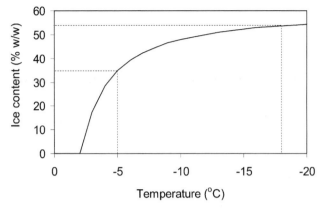

Figure 4.13 *Ice content of a typical ice cream as a function of temperature*

53% by weight (this can be calculated from the phase diagram, as in the example in Chapter 2). The latent heat removed (ΔQ_L) is therefore

$$\Delta Q_L = mL = 1.0 \times 0.53 \times 330 \times 10^3 = 1.7 \times 10^5 \text{ J} \qquad (4.2)$$

Thus, approximately five times more heat must be removed to freeze the water than to cool the mix down.

As the mix passes along the barrel, its temperature decreases, and its ice content increases. As we have seen in Chapter 2, both of these cause the viscosity of the mix to increase: the viscosity of the sugar solution increases as the temperature decreases and the solute concentration and the viscosity of the suspension increase as the volume fraction of ice increases. Figure 4.13 shows the ice content (*i.e.* the mass of ice as a percentage of the total mass) of a typical ice cream as a function of temperature between 0 and $-20\,°C$. This is known as the ice curve. The ice content of this formulation is 35% when it leaves the factory freezer at $-5\,°C$ and 54% at a typical storage temperature of $-18\,°C$.

The increase in viscosity due to ice formation begins about one-third of the distance along the barrel. Increasing the viscosity also affects the whippability of the mix, so that most of the small air bubbles are formed in the final two-thirds of the barrel. However, this also means that the mix becomes increasingly hard to beat; and much more energy input is required to rotate the dasher. This extra energy is dissipated in the mix as heat, which must be removed. Eventually there comes a point when the energy input through the dasher equals the energy removed as heat by the refrigerant, *i.e.* the process becomes self-limiting. For this reason, the lowest outlet temperature that can be achieved in a conventional ice cream freezer is about -5 or $-6\,°C$, and about half of the cooling capacity is used to remove heat generated in this way.

The factory freezer is the only place in the entire ice cream manufacturing and distribution chain where new ice crystals are nucleated. It is therefore very important to form many crystals at this stage, since recrystallization during subsequent processing and distribution leads to a reduction their number (and an increase in their mean size). The total number of ice crystals on exit from the factory freezer (for a given outlet temperature) depends mostly on the residence time and dasher speed. Short residence times and slow dasher speeds produce many crystals with a small mean size, whereas long residence times and high rotation speeds result in fewer, larger crystals. This is somewhat counterintuitive. A fast dasher speed means that the time between scrapes is short, and one might therefore expect the ice crystals to be smaller because they do not have much time to grow between scrapes. In fact increasing the dasher speed increases the total amount of energy dissipated as heat, which increases the temperature in the middle of the barrel (even with a constant outlet temperature). Since the ice crystals are exposed to higher temperatures, they undergo faster recrystallization as they pass along the barrel, which ultimately reduces the number of crystals at the outlet.

The outlet temperature controls the total amount of ice (and hence the mean crystal size for a given total number). This determines the viscosity of the ice cream (typically 10 Pa s on exit from the factory freezer) and its ability to retain its shape. A low outlet temperature is required for ice cream that will be shaped by extrusion because it needs to be quite stiff, whereas a warmer temperature may be preferred for ice cream that is dispensed directly into tubs.

As ice cream leaves the factory freezer there is a pressure drop of typically 4 atm (the difference between the pressure inside the barrel and atmospheric pressure), so the air bubbles expand (Figure 4.14).

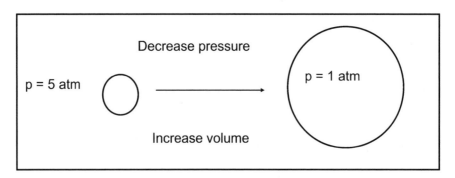

Figure 4.14 *Expansion of air bubbles on leaving the factory freezer*

A calculation using Boyle's law (pressure × volume = constant for a fixed amount of gas at constant temperature) shows that the air takes up about $\frac{1}{6}$ of the volume of mix inside the factory freezer, but about half of the volume of the ice cream on exit.

The overrun was defined in equation 2.4. Since the density (ρ) is the mass (m) divided by the volume (V), this can be written as

$$\text{overrun} = \frac{(m_{\text{ic}} / \rho_{\text{ic}}) - (m_{\text{mix}} / \rho_{\text{mix}})}{(m_{\text{mix}} / \rho_{\text{mix}})} \times 100 \qquad (4.3)$$

where the subscripts refer to ice cream and mix. Air does not add any significant mass, *i.e.*

$$m_{\text{ic}} \cong m_{\text{mix}} \qquad (4.4)$$

Therefore, the overrun can be calculated from the densities of the ice cream and the unaerated mix:

$$\text{overrun} = \frac{(1/\rho_{\text{ic}}) - (1/\rho_{\text{mix}})}{(1/\rho_{\text{mix}})} \times 100 = \left(\frac{\rho_{\text{mix}}}{\rho_{\text{ic}}} - 1\right) \times 100 \qquad (4.5)$$

The mix density is simply determined by weighing a known volume of mix. The density, and hence the overrun of ice cream as it leaves the factory freezer, is measured using an overrun cup. The cup, whose volume is known, is carefully filled to over-flowing with ice cream, ensuring that no air gaps are left, and that the ice cream is not compressed. The ice cream is then levelled off at the top with a knife. The mass of the ice cream is determined by weighing the full cup, and subtracting the known mass of the empty cup.

HARDENING

When ice cream leaves the factory freezer at about $-5\,^\circ\text{C}$, its ice content is only about half that at a typical serving temperature of $-18\,^\circ\text{C}$, so it is very soft. As shown in Chapter 2, the microstructure of dispersed ice crystals and air bubbles is thermodynamically unstable – the system tends towards a state in which the phases are less dispersed. If the ice cream were simply stored at the factory freezer outlet temperature it would deteriorate very quickly. The ice crystals and air bubbles would coarsen: their mean size would increase and their total number would decrease. Since it is not possible to stabilize the microstructure *thermodynamically*, the best that can be achieved is to trap it *kinetically*, *i.e.* to slow down the rate at which coarsening occurs, so that significant deterioration of the microstructure does not take place within the lifetime of the ice cream. Furthermore, it is necessary to combine ice cream with other components, such as cones, chocolate, fruit *etc.*, to assemble particular products. Depending on the product, the ice cream may need to be harder and colder before product assembly can take place, *e.g.*, if it is to be covered in chocolate. For these reasons, the temperature of the ice cream is lowered as quickly as possible after it leaves the factory freezer. This is known as hardening.

Ice cream is hardened in a hardening tunnel, an enclosed chamber into which the ice cream passes on a conveyor belt from the factory freezer. Inside, cold air (typically -30 °C to -45 °C) is blown over the ice cream. The lower the air temperature, and the faster the air flow, the faster heat is removed from the ice cream. Air turbulence also increases the rate of heat transfer. The chamber is enclosed to minimize exchange of cold air inside the system with warm ambient air, and so to reduce the build up of frost that would reduce the efficiency. Cold stores, which are typically about -25 °C, are not suitable for hardening because they are not cold enough and have still air, so they cannot cool the ice cream rapidly enough to minimize recrystallization.

Ice crystals grow during hardening in two ways: by propagation and by recrystallization. Propagation is simply the increase in size of all the ice crystals as water freezes and forms more ice according to the ice curve (Figure 4.13). Recrystallization (coarsening) is the process in which large ice crystals grow at the expense of small ones, so that there is an increase in the mean crystal size, but no change in the total amount of ice. The lower the temperature, the slower the diffusion of the water and sugar molecules in the matrix, and the slower the recrystallization. Thus, to ensure that propagation dominates, and that recrystallization is minimized, hardening should be as rapid as possible. Recrystallization takes place by two mechanisms, Ostwald ripening and accretion. Ostwald ripening dominates at low ice phase volumes, *i.e.* at the beginning of hardening. As the phase volume of the ice increases due to propagation, accretion becomes more important since the distance between the crystals is smaller, so that two crystals in close proximity come into contact as they grow. The increase in the ice phase volume during hardening also causes further freeze-concentration of the matrix. New crystals are not formed during hardening because the driving force for crystallization provided by the cooling is not large enough for nucleation to take place.

Figure 4.15 shows both the mean ice crystal size and total number of ice crystals as a function of ice content (which is directly related to temperature by the ice curve) in a standard ice cream. Samples were taken from the centre of a block of ice cream, and measured

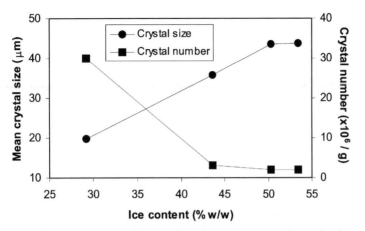

Figure 4.15 *Changes in ice crystal size and number in ice cream during hardening*
(Reprinted from 'Influence of Freezing Conditions on Ice Crystallisation
in Ice Cream',[1] Copyright 1999, with permission from Elsevier)

on exit from the factory freezer and during hardening. The total
number decreases, which shows that recrystallization is taking place
during hardening, and the ice crystal size increases, both because of
propagation and recrystallization.

The air cells also coarsen after leaving the factory freezer, by
disproportionation and coalescence. Figure 4.16 shows the air bubble

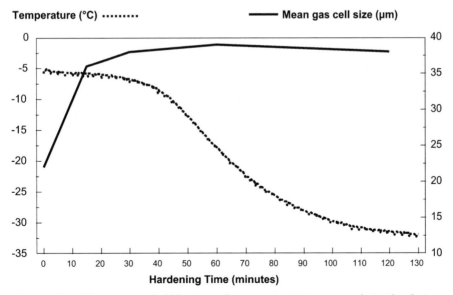

Figure 4.16 *Changes in air bubble size and temperature in ice cream during hardening*
(Reproduced with permission from ref. 2)

size and the temperature in a block of ice cream as a function of time during hardening. This shows that much of the coarsening takes place during the early stage of hardening when the temperature is relatively high. This is because at lower temperatures, the matrix becomes more solid-like, so the air cells are less subject to coalescence, and diffusion is slower, so there is less disproportionation. Therefore, hardening should be rapid in order to minimize the coarsening of the air bubbles.

The length of time that the ice cream needs to spend inside the hardening tunnel depends on several factors, namely the overrun, the formulation, the outlet temperature, the size of the product and the amount of packaging. The greater the overrun, the lower the heat capacity and the thermal conductivity of the ice cream. A lower heat capacity reduces the amount of heat that has to be removed, whereas a lower thermal conductivity reduces the rate at which it is removed. Thus these two effects are opposed, but the heat capacity is the larger effect, so the higher the overrun, the shorter the hardening time. The formulation, particularly the amount of sugar, determines how the ice content varies with temperature. Formulations that have a large ice content at the exit temperature will cool faster than those with a lower ice content, because less ice is formed during hardening so less latent heat has to be removed. Similarly, if the outlet temperature is low, the ice cream is colder to begin with, and less sensible heat has to be removed. Large products take longer to harden than small ones because heat transfer from the centre of the product to the outside is slow. Small products, such as cups, cones and stick products take about 15–20 min for the centre to reach $-20\,^\circ\text{C}$ in a standard hardening tunnel, whereas a two litre tub of ice cream requires about 2 hours. The mean ice crystal size after hardening is larger at the centre of a block of ice cream than at the edge because the centre of the product cools more slowly than the outside, and therefore undergoes greater recrystallization. Similarly, packaged products harden more slowly than unpackaged ones.

A different method is used to measure the overrun after hardening since forcing hard ice cream into an overrun cup would change its volume. A volume displacement chamber is used, *i.e.* a large container with an overflow spout filled with cold water up to the bottom of the spout (Figure 4.17). The ice cream is weighed, and then held with tongs just below the surface of the cold water. The volume that flows out of the spout is measured. Subtracting the volume of the portion of the tongs that was below the surface gives the volume of the ice cream and hence the density and overrun. (Experiment 14 in Chapter 8 describes another method of measuring the overrun of hardened ice cream.)

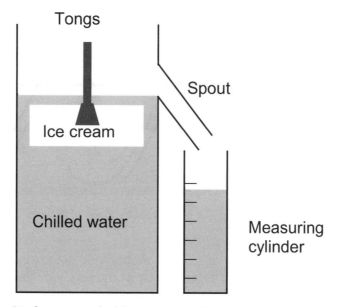

Figure 4.17 *Displacement method for measuring volume*

The volume of a given mass of ice is approximately 8% larger than the volume of the same mass of water. Therefore, the overrun of ice cream changes as its ice content changes, even when the amount of air it contains is constant. For example, one litre of a typical ice cream mix weighs 1.1 kg, and contains 55% ice (*i.e.* 600 g) at −18 °C. The difference in volume between 600 g of ice and 600 g of water is $0.08 \times 600 \approx 50$ cm³. If the ice cream is aerated with 1 l of air (*i.e.* nominally 100% overrun), the volume of the ice cream is 1 l mix + 1 l air + 0.05 l due to change of volume of water on freezing. Therefore the overrun is actually (2.05 − 1) / 1 × 100 = 105%, not 100%. This is a relatively small effect for ice cream, and is often ignored. However, the volume of *frozen* unaerated mix can be measured by the displacement method and used in the overrun calculation to eliminate this error.

Low Temperature Extrusion

Recently a new freezing method, low temperature extrusion, has been developed to overcome the self-limiting nature of the factory freezer, and to avoid the need for hardening. Ice cream from the factory freezer at about −5 °C is passed through a screw extruder with refrigerated walls and cooled to about −15 °C (Figure 4.18). The extruder applies a higher shear stress (and lower shear rate) to the ice cream than a

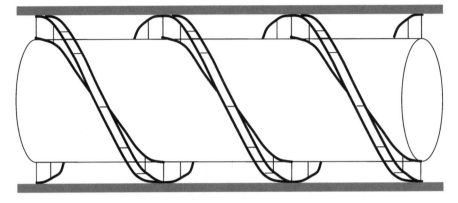

Figure 4.18   *Screw extruder for low temperature extrusion of ice cream*

conventional factory freezer, which means that it can operate at low temperatures and high ice contents despite the very high viscosity of the ice cream. Low temperature extrusion produces smaller air bubbles and ice crystals than conventional freezing, and hence can also produce improvements in the texture. The finer microstructure arises from a combination of the lower temperature at which it leaves the extruder (so that there is less coarsening of the ice crystals and air bubbles) and increased partial coalescence of the fat due to the higher shear stress. This can necessitate changes in the formulation. Because ice cream leaves the extruder at a low temperature, hardening is not required and it can be taken straight to the cold store. However, the higher viscosity of the ice cream makes it harder to mix in sauces or inclusions, and to extrude the ice cream into packaging.

WATER ICES

Water ices can be produced using the ice cream process, but since they usually do not contain fat and often are not aerated, some steps are not required. The ingredients are dosed and mixed, and the mix is then pasteurized. Homogenization and ageing are unnecessary, because of the absence of fat. The mix is frozen in the factory freezer, usually without the injection of air. In water ice mixes, the ice phase volume changes rapidly with temperature. Therefore, a small fluctuation in temperature can cause a large change in ice phase volume. If the amount of ice becomes too high it can form large solid lumps in the barrel, which can damage the scraper blades, prevent the dasher from rotating and can cause the motor to burn out. This is known as icing up. Using an open dasher, and hence a large volume of mix, gives a buffer against such fluctuations and reduces the likelihood of icing up. For aerated water ice products, such as sorbets, a barrel pressure of 2–3 atm is used to achieve the required overrun. On exit from the factory freezer, a water ice slush is produced, with an ice phase volume that is determined by the outlet temperature. The size and shape of the ice crystals are similar to those in ice cream. Some simple water ice products are produced by quiescent freezing, *i.e.* by simply freezing mix in a mould in a bath of cold liquid without passing through the factory freezer.

REFERENCES

1. A.B. Russell, P.E. Cheney and S.D. Wantling, *J. Food Eng.* 1999, **39**, 179.
2. S. Turan, M. Kirkland and R. Bee, in 'Food Emulsions and Foams: Interfaces, Interactions and Stability', ed. E. Dickinson and J.M. Rodriguez Patino, The Royal Society of Chemistry, Cambridge, 1999, Special Publication No. 227, p. 151.

FURTHER READING

R.T. Marshall, H.D. Goff and R.W. Hartel, 'Ice Cream', 6th Edition, Chapman & Hall, New York, 2003, Chapters 6–9.

Chapter 5

Product Assembly

The factory process by which ice cream is produced has been described in Chapter 4. This is not the end of the story, however, since many products consist of more than just ice cream or water ice. The ice cream may be held in a cone or on a stick, it may be coated, for example in chocolate, and it may contain inclusions, such as pieces of fruit, nut or biscuit. Packaging is also an important part of most ice cream products. As indicated in Figure 4.1, products are assembled and packaged after the ice cream leaves the factory freezer or after hardening, depending on the nature of the product. Figure 5.1 shows a schematic diagram of the main steps in product assembly and packaging.

Ice cream products can generally be classified in three groups.

- Tubs or desserts that are bought in supermarkets and are eaten at home, providing several servings.
- Soft ice cream, such as *Mr Whippy*, that is made in a small ice cream freezer, for example in an ice cream van, and eaten immediately.
- Single-serve products, such as cones, ice creams or ice lollies on sticks and choc ices, that are bought singly from a shop or ice cream van, or in multi-packs from a supermarket. These are often described as 'impulse' (because they are often bought on impulse) or 'novelty' products (because manufacturers usually introduce new products or variants each year to maintain consumer interest through novelty).

Tubs and soft ice creams do not require substantial product assembly. Tubs are filled with ice cream straight from the factory freezer and then hardened. Inclusions may be added at the point of filling, using a fruit feeder. This consists of a hopper for the inclusions; a means for

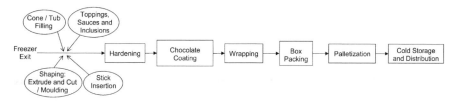

Figure 5.1 *Schematic diagram of the product assembly and packaging process*

ensuring that the correct amount is added, such as an auger; a rotor or plunger for inserting the inclusions; and a chamber where they are mixed with the ice cream. Sauces or ripples are added by co-extrusion. Since inclusions, toppings and sauces are added after pasteurization, it is essential that they do not introduce micro-organisms. Soft ice cream is produced in a small-scale ice cream freezer from a pre-made mix, following the process described in Chapter 4, except that it is eaten immediately after freezing and therefore does not need to be hardened. Impulse products and desserts are more complex, and are produced by a number of different processes (according to the nature of the product). These are described below, but first it is necessary to look in more detail at the most important and widely used component of ice cream products, other than the ice cream itself, namely chocolate and couverture.

PHYSICAL PROPERTIES OF CHOCOLATE AND COUVERTURE

Chocolate and couvertures are dispersions of small (10–25 μm) particles of sugar, and, depending on the type of chocolate, cocoa powder solids and non-fat milk solids, in a continuous fat phase. A layer of emulsifier (usually lecithin) is strongly adsorbed to the surface of the sugar particles. Chocolates and couvertures for ice cream products generally have higher fat contents (40–60%) than their normal ambient counterparts (28–35%). This is because chocolate on ice cream products is eaten at lower temperatures than ambient chocolate. An ambient chocolate would be very hard at $-18\,°C$, and would have a waxy texture when eaten in conjunction with ice cream, because the mouth would be too cold to melt the fat in the chocolate.

The Casson equation (equation 5.1) is often used to describe the rheology of liquid chocolates and couvertures. The relation between the shear stress (σ) and the shear rate ($\dot{\gamma}$) depends on two parameters: the yield stress, σ_y, *i.e.* the shear stress required to initiate flow, and the plastic viscosity, η_{pl}, *i.e.* the viscosity after yield.

$$\sigma^{0.5} = \sigma_y^{0.5} + \eta_{pl}(\dot{\gamma})^{0.5} \qquad\qquad (5.1)$$

Several factors affect the rheology: the formulation (amount and type of fat, the emulsifier, milk, and water contents), the size distribution of the solid particles, and the temperature. The viscosity is usually modified to suit a particular process by changing the fat content. For example, molten ambient chocolate is quite viscous, so coating an ice cream product with it would produce a layer that is too thick. Increasing the fat content reduces its plastic viscosity and yield stress, so a thinner coating can be produced. Adding lecithin also lowers the viscosity, by reducing the friction between the particles. Water is not deliberately added to most chocolate formulations, but small amounts, for example residual moisture from the milk solids or water picked up from ice cream during coating, can hydrate the sugar particles. This strengthens the interactions between them and dramatically increases the viscosity. Lecithin helps to counteract this by coating the sugar particles so that the water cannot stick them together. The viscosity and other properties of couvertures can be adjusted by choosing appropriate types and amounts of fats.

The main methods for coating ice cream products with chocolate or couverture are spraying (for cones), dipping (for stick products) and enrobing (for bars). To achieve an even coating of the correct thickness without any holes, each of these methods imposes specific requirements on the rheology of liquid chocolate. For example, the yield stress affects drainage of the coating. A high yield stress is necessary to prevent the formation of 'feet' on enrobed products, and a low yield stress is needed to produce an even coating on dipped products. The plastic viscosity determines the behaviour during pumping and spraying, where the chocolate or coating is subjected to relatively high shear rates.

Similarly, there are requirements on the mechanical properties of the chocolate after it has set. For example, chocolate coatings on dipped products should not to be so brittle that they fracture during packaging, but must be brittle enough to crack when bitten in order to provide texture contrast with the ice cream. The mechanical properties of solid chocolate are determined by the crystallization of the fat. This is a complex area, because fats are usually mixtures of different triglycerides. Cocoa butter consists mostly of three triglycerides: 20% POP, 40% POS and 25% SOS, where palmitic (P), oleic (O) and stearic (S) denote the fatty acids. Cocoa butter can exist in six different crystalline forms (polymorphs). Ambient chocolate has to undergo a carefully designed cooling process (called tempering), in order to obtain a stable polymorph. If it is not cooled properly, a transition between two different polymorphs can occur, which leads to the formation of a dull grey surface, a defect known as bloom. However, chocolate on ice cream does not need to be tempered. It is rapidly cooled and stored at a low temperature, which prevents the formation of bloom.

When it first solidifies, chocolate has a plastic or leathery texture, due to incomplete solidification of the fat. It slowly solidifies, taking minutes or hours to become brittle. There are two important process parameters associated with fat crystallization. The drying time is defined as the time at which the coating no longer leaves a smear on touching a wrapper, and is typically one minute. The brittleness time is defined as the time at which the coating first develops an audible crack on biting, and is typically 2–3 min (however, the chocolate continues to become more brittle after the brittleness time). Any increase in the complexity of the triglyceride composition (*e.g.* by adding milk fat) makes it more difficult for the fat to crystallize. This generally results in a decrease in crystallization rate, *i.e.* an increase in the drying and brittleness times. A typical ice cream couverture starts to crystallize at lower temperatures, crystallizes over a narrower temperature range and reaches its maximum brittleness much more quickly than chocolate.

CONES AND SANDWICHES

Cones are complex products, consisting of a wafer cone, internally coated with chocolate or couverture, filled with ice cream and topped with additions, such as nuts or chocolate pieces. Their assembly requires several steps (Figure 5.2).

Wafers are made by depositing batter mix onto hot baking plates. The baked wafer sheets are removed from the plates while still soft and

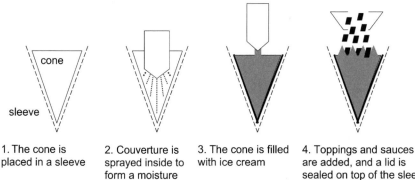

1. The cone is placed in a sleeve

2. Couverture is sprayed inside to form a moisture barrier

3. The cone is filled with ice cream

4. Toppings and sauces are added, and a lid is sealed on top of the sleeve

Figure 5.2 *Assembly of a cone product*

Figure 5.3   *Spraying the inside of cones with couverture*

transferred onto rollers that shape them into cones. The cones are removed from the rollers, allowed to cool and placed in packaging sleeves. To prevent the wafer from becoming soggy by absorbing water from the ice cream, the inside of the cone is sprayed with couverture to form a moisture barrier (chocolate is generally too viscous for spraying). Figure 5.3 shows cones being sprayed.

To be effective, the sprayed couverture must form a continuous layer, completely covering the interior and rim of the cone. The coating should have a high yield stress to minimize drainage to the bottom of the cone. Drainage is difficult to eliminate altogether, which is why some cone products have a plug of couverture at the bottom of the cone. The couverture should set quickly so that furrows are not formed in the coating when it is filled with ice cream. It should also not be too brittle so that it does not crack if subjected to stress. A couverture with a relatively high amount (*ca.* 65%) of coconut oil meets these criteria.

A few seconds after spraying, ice cream is extruded from a nozzle into the cone. The top of the ice cream is often fluted to enhance its appearance, so the ice cream must be viscous enough for it to retain its shape after extrusion. Sauces or inclusions are then dispensed to provide an attractive top to the product, and the top is sealed with a lid. Finally,

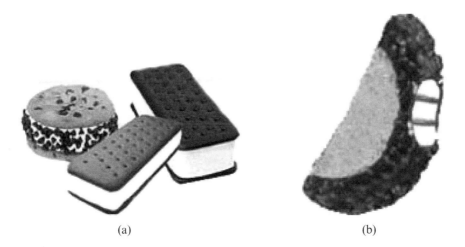

(a)   (b)

Figure 5.4   (a) *Examples of sandwich products* (b) *a taco*

the cones are hardened and packed into boxes that are stacked onto pallets.

Sandwiches consist of one or more flavours of ice cream between two biscuits or wafers, which may have been covered in chocolate (Figure 5.4a). Biscuits are made by mixing flour, sugar and fat, shaping and then baking. The sandwich can be assembled layer by layer, *i.e.* the first biscuit or wafer is covered with ice cream, which in turn is covered with a second biscuit or wafer. Alternatively, the ice cream can be extruded between two biscuits. The product is then hardened, dipped in chocolate or couverture if required, and packaged.

Tacos are a type of sandwich product consisting of a semi-circular shaped wafer that contains ice cream and sauce (Figure 5.4b). The process for producing tacos is shown in Figure 5.5. Circular wafers are produced as described above. The ice cream and sauce are co-extruded and cut in a semi-circular shape, before being hardened and enrobed in couverture. While the wafer is hot it is flexible, the enrobed ice cream is pushed onto it, causing it to fold around the ice cream. Some of the couverture melts when it touches the wafer, and bonds the wafer to the ice cream when it sets again. The couverture also forms a moisture barrier between the wafer and the ice cream. The taco is allowed to cool before it is dipped in chocolate or couverture. As it is dipped, the taco is rotated so that only the outer circular rim is coated. Finally, the taco is packaged.

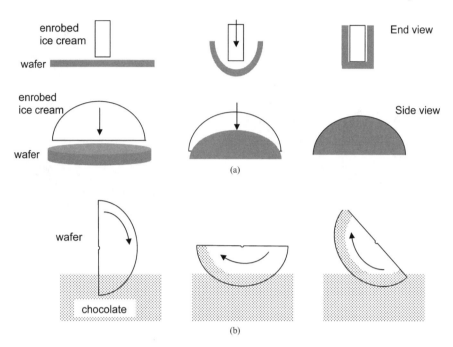

Figure 5.5 *Schematic diagrams of taco production:* (a) *end and side views showing how the wafer is folded over the ice cream and* (b) *dipping of the rim*

STICK PRODUCTS, BARS AND TUBES

The use of wooden sticks for ice cream and water ice products dates back to 1921. For many years, most stick products were ice lollies, and most bars were couverture coated ice creams ('choc ices'). Today the variety is much greater and ice cream, water ice and milk ice are sold in a number of different formats. For example, premium chocolate-coated ice creams are now often produced as stick products, and water ices are sold in tubes. Two main processes, moulding and 'extrude and cut' are used to produce stick products. Tubes use the package as the mould and bars are made by extrude and cut. Bars and stick products are often coated with chocolate or couverture, and sometimes also dry pieces such as nuts.

Moulding

Moulded stick products may be made from water ice, milk ice, ice cream, or any combination of these. Moulded products are quiescently frozen, often using a rotating table (Figure 5.6). The moulds are made

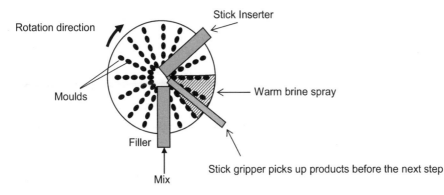

Figure 5.6 *Moulding process on a rotating table, viewed from above*

from metal or rubber. Metal moulds are robust, but they can only be used for products with simple, straight or tapered shapes and a low level of surface detail. It would not be possible to extract other shaped products from the mould without breaking them. Rubber moulds are flexible, and allow more complex three-dimensional shapes to be produced, for example cartoon characters. However, sharp features, such as ears, are still liable to be broken during demoulding.

The moulds are immersed in a tank underneath the table, which contains brine (calcium chloride solution) or a mixture of water and ethylene glycol at about $-40\,°C$. Unfrozen mix (or sometimes mix that has been partially slush frozen to an ice phase volume of about 10%) is pumped to a filler that doses it into the moulds. The mix begins to freeze from the outside towards the centre. Sticks, which are generally wooden and need to be smooth, flat and free from splintering, are inserted when the product is hard enough for them to stand up, but still soft enough to allow insertion without effort. Since freezing starts on the mould, the ice crystals grow towards the centre. As a result they are large, elongated and aligned, and are visible to the naked eye when you bite into an ice lolly. Lines can also sometimes be seen where the growing ice fronts from different sides of the mould have met. Ice lollies usually have a high ice content, which together with the very large ice crystals results in a hard texture.

When the product is completely frozen, warm brine is sprayed onto the underside of the moulds. The temperature of the brine and the time for which it is applied are chosen so that just enough heat is transferred to the mould to release the product from the mould. Too much heat melts the outside of the product, affecting its detail and shape, and leaves mix in the mould. Too little heat results in poor extraction, for example breaking sticks or leaving product behind in the mould. The ease of de-moulding depends on the shape of the product. For example, it is easier to de-mould tapered products than rectangular ones. The moulds are then cleaned ready for the next filling. Hardening is not required since the product is very cold when it is removed from the mould. The product is ready for further processing and packaging. For example, a dipping station may be placed after the rotating table to coat the product in chocolate. One advantage of using a rotating table is that the arms can be moved around the table, which allows the timings of the steps to be varied easily. For example, increasing the angle between the filling and stick insertion arms means that there is more time for freezing before the stick is inserted. This might be used when changing between mixes with different formulations and hence different ice curves.

Tube products use their own package as the mould. The tube is dispensed from a stack into a holder where it is held vertically by its rim. The liquid mix is dispensed into the tube, and the lid is picked up from a stack, placed on top by a vacuum sucker and heat-sealed onto the rim. The filled and sealed tubes are then frozen on a conveyor belt in the hardening tunnel.

A variation on the moulding process is to suck out unfrozen material from the centre of the mould shortly after filling, to leave a frozen outer shell. This can be refilled with another mix to create a two-component product, such as a 'split' with an ice cream core and a water ice shell. This is known as 'fill and suck' (Figure 5.7).

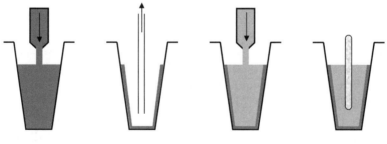

1. The mould, which is immersed in cold liquid, is filled with the first mix

2. Some of the mix freezes on the walls, and the remaining unfrozen liquid in the centre is sucked out

3. The shell is then filled with the second mix

4. As the second mix freezes, a stick is inserted

Figure 5.7 *Fill and suck process*

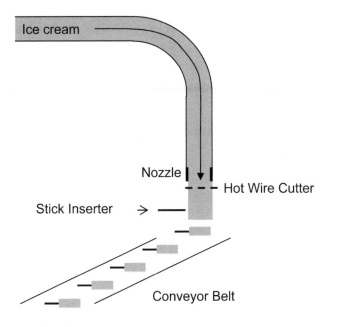

Figure 5.8 *Extrude and cut process*

Extrude and Cut

Figure 5.8 shows the extrude and cut process. Ice cream from the factory freezer is extruded though a shaped nozzle onto a plate conveyor. Stick products are usually extruded vertically downwards, and bars are extruded horizontally. A heated wire (vertical extrusion) or knife (horizontal extrusion) cuts it into portions of the required size. Sticks are inserted at the nozzle prior to cutting if required. The formulation and processing conditions (particularly the outlet temperature) are carefully chosen to produce ice cream with the correct rheological properties on extrusion. The ice cream must be sufficiently hard that it does not flow after it has been cut, so that it does not slump before hardening and the stick stays in place. However, it must not be so hard that it is difficult to insert the stick or that cutting produces ragged edges.

Clever nozzle design allows the production of complex two-dimensional shapes, and co-extrusion of different flavours of ice cream, or sauces, enables detail to be added. The extrude and cut process can therefore produce products that cannot be made by moulding. Figure 5.9 shows two examples of these. Spirals of ice cream can be produced with a rotating nozzle: two ice cream streams rotate around the centre, through which a further ice cream stream is flowing.

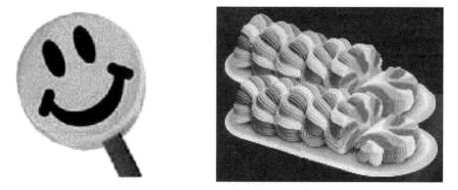

Figure 5.9 *Two examples of shaped products produced by extrusion*

The conveyor carries the products through the hardening tunnel. During hardening the ice cream sticks to the conveyor plate. To release it a hammer strikes the underside of the plate when the product leaves the hardening tunnel as it is picked up by a stick gripper. The stick gripper consists of sets of tongs mounted on a carriage frame. It is mechanically complex (especially on production lines that require the product to be turned through 90°) and must be correctly positioned so that the products are picked up, transferred and then accurately placed in subsequent steps in the product assembly process, such as dipping and wrapping.

Dipping

Stick products can be covered in any type of quick-setting coating, such as chocolate or couverture, by dipping (Figure 5.10). The chocolate or couverture (which may also contain solid pieces, such as nuts) is held in a heated tank at about 45 °C and is continuously stirred to prevent the particles from settling out. The stick gripper transfers the products to the dipping tank, where they are lowered into the liquid for a certain time, and then removed. The ice cream is cold enough to set the chocolate without melting itself. The liquid level is set so that the products are immersed to the correct dipping depth. The dipping time, product temperature and viscosity of the liquid determine the thickness of the coating deposited on the product. Since chocolate and couverture are expensive, it is important to control the process tightly. If the dipping time is too short, the ice cream too warm, or the viscosity too low, the coating may be incomplete. The reverse of these can result in a coating that is too thick.

A thick coating is usually required for premium chocolate-covered stick products, so chocolate with reasonably high plastic viscosity and

Figure 5.10 *A stick product that has just been coated by dipping in chocolate*

yield stress is used. This is achieved with *ca.* 45% cocoa butter. Dipping couvertures are usually considerably less viscous than dipping chocolates because thinner coatings are required. A typical dipping couverture contains 55–60% coconut oil. Coconut oil couvertures are brittle, so up to 10% of a liquid oil, such as sunflower or soybean, or milk fat may be added to avoid cracking during packaging.

After dipping, the products pass over a wire above a drip tray to remove and recover drips of coating that form at the bottom of the product. They are then conveyed to the wrapping machine, allowing time for the coating to cool and set hard. The coating must set rapidly to avoid smearing on the inside of the wrapper, but not become so brittle that cracking occurs during packaging. The colder the ice cream, the faster the coating sets.

Dipping can also be used to coat ice cream with fruit ice. More heat must be removed from fruit ice than from chocolate because it has a lower freezing point and a higher heat capacity. Ice cream at about $-25\,^{\circ}\text{C}$ from the hardening tunnel cannot itself provide enough cooling to freeze the fruit ice. Instead the ice cream is dipped in liquid nitrogen (at $-196\,^{\circ}\text{C}$) immediately before and after dipping in the fruit mix. Quite a thick fruit coating is often required, so more than one dipping

Ice cream on stick from   Coated product
extrude and cut process

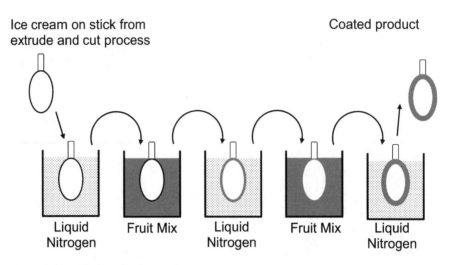

Liquid   Fruit Mix   Liquid   Fruit Mix   Liquid
Nitrogen   Nitrogen   Nitrogen

Figure 5.11   *The liquid nitrogen dipping process*

step may be used. Figure 5.11 shows this process. Since liquid nitrogen is very cold, freezing is nucleation dominated and, thus, many small ice crystals are formed. The small ice crystals and the high solids content of the fruit mix mean that the fruit ice coating is soft; it would not be possible to produce this sort of product by fill and suck.

Liquid nitrogen is also used to freeze another type of product. Small spheres of ice cream or water ice can be formed by dripping mix through nozzles into a tank of liquid nitrogen. The droplet size is controlled by the size of the nozzle. When the droplets fall into the liquid nitrogen, they freeze rapidly. This method enables a large number of frozen drops to be produced quickly. The droplets are removed from the tank, and packaged in tubs. Figure 5.12 shows an example of this type of product.

Enrobing

Unlike stick products, ice cream bars cannot easily be picked up for dipping, so a different process, called enrobing, is used (Figure 5.13). The product enters the enrober from the hardening tunnel along a mesh conveyor belt and passes through one or more waterfalls of chocolate or couverture, known as curtains. Next, an air knife blows off the excess coating, which drains through the mesh conveyor. Finally the bottom of the bar is immersed in liquid chocolate to ensure that the underside is also coated.

The rheology of the coating is critical, both at low shear rates as the bars pass through the curtain, and at higher shear rates when the air

Figure 5.12 *Solero Shots are droplets of water ice that are frozen in liquid nitrogen*

Direction of motion of ice cream bars ⟶

Curtains   Air knife   Immersion of bottoms

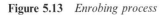

Figure 5.13 *Enrobing process*

blower removes the excess coating. Enrobing chocolates and couvertures have higher plastic viscosities and yield stresses than dipping ones,
because greater pick-up of the coating is required. Enrobing chocolates

and couvertures typically contain about 50% cocoa butter and coconut oil respectively. Whole milk powder and/or cocoa may be added to reduce the brittleness.

Dry Coating

Dry coating is the addition of dry components, such as nuts and biscuit crumbs to the outer surface of a product. The surface of the product must be tacky so that the dry coating will adhere. This can be achieved either by melting a thin layer on the surface of a moulded stick product, for example by increasing the time or temperature of the warm brine spray during de-moulding, or by using a dipped product before the liquid coating has set. The product is lowered into the dry coating unit and the dry ingredients are thrown onto its surface by rotating blades or blowers, in order to achieve a uniform distribution.

DESSERTS

There are many different kinds of dessert product and processes to produce them, so manufacturing lines must be flexible. Desserts usually contain more than one flavour of ice cream and/or water ice, so they are commonly made by co-extrusion of the components. The co-extruded components (*e.g.* ice cream and sauce) should have similar viscosities so that their flows can be balanced and the correct product shape produced. If necessary, the viscosity of a sauce can be increased by first passing it through a factory freezer. A closed dasher is used as the throughput is very low and a short residence time is required. This also has the advantage that the sauce then has approximately the same

Figure 5.14   *A co-extruded log product*

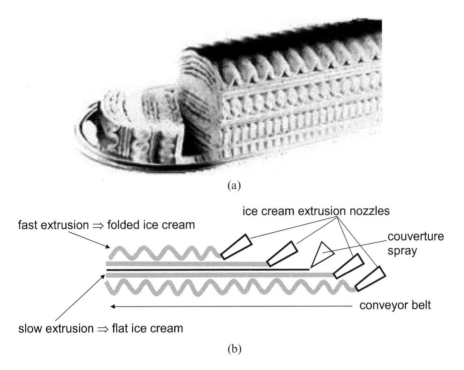

(a)

(b)

Figure 5.15 (a) *Vienetta;* (b) *diagram of its manufacture by extrusion*

temperature as the ice cream. Extruded logs are produced using the same process as for extruded bar products, except on a larger scale. Since they are larger (typically 0.5–1.5 l), desserts require longer hardening times than impulse products. Figure 5.14 shows an example of a co-extruded log dessert.

Other more complex desserts require clever process design. A good example of this is *Vienetta* (Figure 5.15a), one of the most popular ice cream desserts, which consists of layers of ice cream and couverture. *Vienetta* is produced by extruding ice cream through a series of nozzles onto a moving conveyor. Thin layers of couverture are sprayed between the layers of ice cream. The ice cream is extruded faster than the conveyor so that it folds up on itself, creating the fluted layers (Figure 5.15b). The flow behaviour of the ice cream after extrusion is crucial. It must flow and fold over itself as the layers are formed, and then retain its shape as other layers are extruded on top. If the ice cream is not stiff enough the lower layers will not be able to support the complete structure, and the flutes will collapse. The formulation and extrusion conditions are chosen accordingly.

PACKAGING

The packaging is an essential part of an ice cream product, as it both protects and helps to sell it. Packaging may take place during and/or after product assembly, depending on the product. There may be more than one layer of packaging – for example wrapped stick products are packaged in boxes with, for example, 24 or 48 products per box (the second layer), which are then stacked onto a pallet and shrink-wrapped (the third layer). The second and third layers are removed after distribution, so that the consumer usually only sees the first layer. The choice of material and structure of packaging is determined by several considerations.

- Functionality, *i.e.* to protect and preserve the product through storage and distribution.
- Customer appeal, *i.e.* to enhance the appearance of the product.
- Safety: there are special requirements for materials that come into contact with food. Packaging must be manufactured in hygienic conditions and must not transfer toxic substances to the product.
- Labelling: to meet marketing, logistical and legal requirements, packs must display information such as the list of ingredients. Cases used for secondary packaging display information for controlling the flow and storage of products in the distribution chain.
- Recycleability: packs should be made from a single type of material to enable effective recycling.
- Production and distribution costs: the number of products that can be fitted onto a pallet for storage and transportation has a significant effect on the final cost of the product.

Examples of the four main types of packaging are shown in Figure 5.16.

Bars and stick products are flow-wrapped. The printed wrapping, which may be coated paper, orientated polypropylene or metallized film, is supplied as a long roll. The wrapping is unwound from the roll and made into a continuous tubular sleeve (Figure 5.17a). Products are placed in the sleeve (Figure 5.17b), which is heat or pressure sealed along its seam. Finally, the tube is cut across to separate the products and the ends of the wrapping are sealed.

Cone sleeves are made from a laminate of aluminium foil and paper. The top of the sleeve is crimped down over a cardboard disc to form a lid after product assembly is completed. Tubes are made from cardboard coated on both sides with polyethylene. The base is heat-sealed to make it leak-proof and the top is rolled to make a rim, onto which the lid is heat-sealed. Plastic tubs and containers are used for ice cream and

Figure 5.16 *Examples of common types of ice cream packaging:* (a) *flow wrap,* (b) *a cone sleeve and a tube,* (c) *tubs and* (d) *a dessert container*

desserts. A range of different sizes are used: single serving plastic cups (50 to 150 ml), take home packs, usually 1 or 2 l in Europe, or ½ gallon or one gallon (3.79 l) in North America, and large packs (up to 10 l) for restaurants and scooping parlours. Desserts are often placed in plastic containers to protect their structure and then individually packaged in boxes.

COLD STORAGE AND DISTRIBUTION

Once the product has been packaged, the manufacturing process is complete. The palletized products are sent to the factory cold store, which typically operates at about $-25\,^{\circ}$C. As well as acting as a hold/release store, the cold store ensures that the product is at $-25\,^{\circ}$C prior to distribution. If the product enters the distribution system at or below $-25\,^{\circ}$C it can undergo some exposure to warmer temperatures before it reaches a temperature at which it deteriorates rapidly. The purpose of the distribution system, from factory cold store, through the retailer's cold store to the shop freezer is to keep the product cold to prevent a decrease in quality due to recrystallization and coarsening. Care must be taken at each stage to ensure that temperature fluctuations are minimized, as repeated increases and decreases in temperature (for example

(a)

(b)

Figure 5.17 *Flow wrapping: (a) the wrap comes off the roll and is shaped into a sleeve; (b) products are placed in the sleeve, which is closed and sealed along its seam*

during transfer from one part of the distribution system to the next) can cause greater recrystallization than storage at a constant temperature. Although the primary difficulty with preserving the quality of ice cream through the distribution system is its temperature sensitivity, ice cream is also sensitive to pressure and mechanical shock, and can absorb odours. The distribution system should also protect ice cream against these if necessary. Factory made ice cream usually needs to have a shelf life of at least 6 months. One reason for this is that ice cream consumption is very seasonal: much more is eaten in summer than in winter. However, it is more efficient to operate a factory all year round. Therefore, products for consumption in the summer may well be produced the previous winter. Provided that the cold storage and distribution chain functions properly, the microstructure that was carefully produced in the factory is preserved so that the ice cream or water ice reaches the consumer in perfect condition!

FURTHER READING

S.T. Beckett, 'The Science of Chocolate', Royal Society of Chemistry, Cambridge, 2000.

R.T. Marshall, H.D. Goff and R.W. Hartel, 'Ice Cream', 6th Edition, Chapman & Hall, New York, 2003, Chapter 12.

Chapter 6

Measuring Ice Cream

In order to develop new formulations, products and processes, for example to create different textures, or make product improvements, scientists need to measure and characterize the microstructure and properties of ice cream. The main measurement techniques used on ice cream can be grouped into four broad classes:

1. Visualization and characterization of microstructure, for example optical or electron microscopy to see ice crystal or air bubble size distributions.
2. Measurements of the response of ice cream to deformation, such as mechanical properties and rheology.
3. Thermal properties, such as the heat capacity and thermal conductivity.
4. Sensory measurements that use the human sensory system to assess texture, flavour and appearance.

Figure 1.1 showed the links between the formulation, the process and the texture. The first step is to understand how the formulation and process affect the microstructure. This requires microscopy techniques to visualize the ice crystals, air bubbles, fat droplets and matrix and image analysis to quantify their sizes, shapes and locations. The next step is to measure the mechanical, rheological and thermal properties and to relate them to the microstructure. The final stage is to relate these physical measurements to the sensory properties. This chapter describes the techniques used to make these measurements.

VISUALIZATION AND CHARACTERIZATION OF MICROSTRUCTURE

Optical and electron microscopy are used to visualize the microstructure of ice cream. Optical microscopy has the advantages of being simple, inexpensive and convenient, but it has limited resolution and the sample preparation alters or destroys some parts of the microstructure. Electron microscopy which uses a beam of electrons to image the sample in the frozen state, and provides a very wide range of magnifications but is more complex and expensive. Electrons are emitted from a source and accelerated towards the sample by a positive voltage. The beam is focused onto the sample using metal apertures and magnetic lenses. There are two different types of electron microscopy, scanning electron microscopy (SEM) and transmission electron microscopy (TEM).

Scanning Electron Microscopy

Scanning electron microscopes work by scanning the electron beam back and forth across the sample in a grid. As the beam hits the sample electrons are knocked off the surface. A detector counts the number of electrons emitted from each spot on the sample and builds up an image: the more electrons emitted, the brighter the corresponding spot on the image. Scanning electron microscopy provides three-dimensional images of the surface microstructure, and can achieve a wide range of magnifications, routinely $\times 10$ to $\times 10000$. It can image the most important length scales in ice cream, from fat droplets (~1 μm) through ice crystals and air bubbles (10–100 μm) to an overall view of the microstructure (>1 mm). SEM is therefore a popular technique for examining the microstructure of ice cream.

Scanning electron microscopes operate under high vacuum, but ice cream at normal processing or storage temperatures would not be stable under these conditions because the ice would sublime and the structure would collapse. However, ice cream is stable below -100 °C: at this temperature the vapour pressure of ice is zero so there is no sublimation, and the matrix is glassy so the ice cream is sufficiently robust to survive the pressure change.

To prepare suitable samples, a small piece (approximately 5×5 mm $\times 10$ mm) is cut from the centre of a block of ice cream with a cold scalpel blade to prevent melting. This is mounted on a sample holder and plunged into liquid nitrogen or nitrogen slush (a mixture of solid and liquid nitrogen at about -200 °C). The very rapid cooling fixes the structure so that it is possible to arrest dynamic events and therefore investigate structural change as a function of time. For example, the data for the air cell size as a function of time during the hardening process that was shown in Figure 4.16 were obtained in this manner.

The sample is next transferred to a vacuum preparation chamber where it is fractured with a knife to expose a clean internal surface. In scanning electron microscopy, contrast arises from topology, *i.e.* differences in the height of the surface of the sample. After fracture, the sample surface is flat (apart from depressions corresponding to the air bubbles). To distinguish the ice crystals the sample is held at just above -100 °C for a short time so that some of the ice sublimes (this is known as etching). The higher the temperature, and the longer the etching, the greater the depth of ice removed. The sample is then cooled back down below -100 °C to stop sublimation. Enough ice should be removed to produce good contrast with the matrix, but not so much that ice crystals disappear altogether. Ice cream is not a good electrical conductor, so when the electron beam hits the surface of the sample, the electrons cannot easily flow away. This would lead to a build up of charge on the surface, which can cause poor imaging. To prevent this, the sample is coated with a thin layer of a good conductor, such as platinum or gold. Finally, the sample is transferred under vacuum from the preparation chamber to the electron microscope where it is imaged at approximately -150°C. This procedure ensures that there are minimal preparation artefacts, so that the images obtained are representative of the actual ice cream microstructure.

Figure 6.1a shows a SEM image of a typical ice cream. The ice crystals appear as irregular shaped objects, often with angular corners or flat sides and typically 50 μm in size. In some cases, it is possible to identify where two or more ice crystals have accreted. The air bubbles are rounder, somewhat larger, and appear both light and dark. Where the fracture is close to the bottom of the bubble, air bubbles form shallow rounded depressions that are bright in the image. Conversely, where the fracture is close to the top, they form deep, nearly spherical holes, which are dark. The apparent air bubble size depends on where it is sliced: if they are sliced near their centre the apparent size is close to the real one. Conversely, if they have been fractured near the top or the bottom of the bubble, the apparent size is smaller than the actual size. (It is possible to correct for this effect statistically.) Some air bubbles are distorted, *i.e.* non-spherical, because of the growth of adjacent ice crystals during hardening. The matrix is the continuous phase between the air bubbles and the ice crystals. The fat droplets (about 1 μm) can be seen on the surface of the air bubbles at higher magnification (Figure 6.1b).

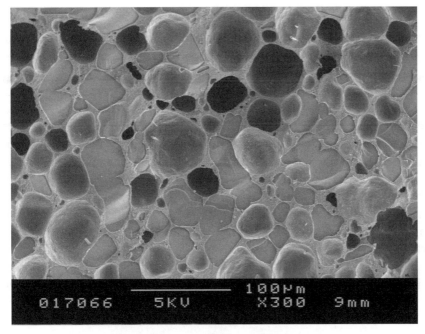

(a)

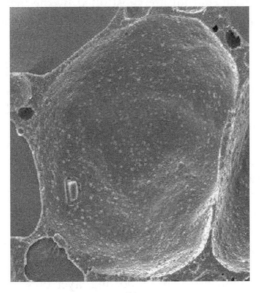

(b)

Figure 6.1 (a) *SEM image showing the ice crystals, air bubbles and matrix in ice cream,* (Reprinted with permission from IOP Publishing Ltd.[1])
(b) *higher magnification image showing the fat droplets on the surface of an air bubble*
(The image is 40 μm wide)

Transmission Electron Microscopy

Transmission electron microscopes work by shining the electron beam through the specimen rather like a slide projector. The beam strikes the sample and a magnetic lens focuses the transmitted portion to form the image. Dark areas of the image correspond to thick or dense parts of the sample through which few electrons are transmitted, and light areas correspond to thin or less dense parts of the sample. TEM can produce higher magnification and higher resolution images than SEM. This means that very fine details of the structure can be visualized, such as fat droplets and casein micelles. The sample must be very thin so that electrons can pass through it. It is possible to stain some components of the microstructure so that they are more visible in the images. Sectioning and staining both necessitate complex sample preparation, for which there are two different methods.

The first, freeze substitution, involves replacing the ice, air and fat with resin. The ice cream is quenched in liquid nitrogen and broken into small pieces (<1 mm^3). These are placed in a vial of methanol at -80 °C. A small amount of a compound such as osmium tetroxide or uranyl acetate is added to stain certain components, such as casein micelles. The ice, air and fat are substituted by the methanol, without destroying the microstructure. The methanol is next replaced by infiltration with resin. The resin is polymerized with ultraviolet light, so that the sample is sufficiently robust to be sliced into very thin sections (about 0.1 µm). Since the resin substitutes for the ice, air and fat, they have similar contrast in the image. Fat can be distinguished by the size of the particles, and ice and air by their shape. Figure 6.2 shows a transmission electron micrograph of the structure of ice cream mix prepared in this manner. Casein micelles are stained dark and are clearly visible near the surface of fat droplets.

The second sample preparation method is freeze–fracture and replication. A small sample is quenched, fractured and etched, similar to the procedure for SEM. The sample is then shadowed with a beam of platinum/carbon or tungsten/tantalum at an angle of 45° to form a layer about 1 nm thick. This converts topological features into variations in electron density. For example, the layer is thin in the shadow of a raised feature (Figure 6.3a). A supporting layer of several nanometres of carbon is then deposited. Carbon is not electron dense so the electron beam passes through it without affecting the image. The replica is floated off the sample, collected on a fine mesh copper grid, and cleaned. Replicas prepared in this manner contain fine microstructural detail, and are stable in the electron microscope. Figure 6.3b, which

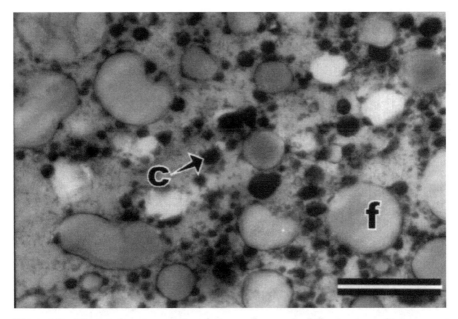

Figure 6.2 *Thin section transmission electron micrograph of the structure of ice cream at high magnification, showing fat droplets (marked f) and casein micelles* (c) (Reprinted from 'A Study of Fat and Air Structures in Ice cream',[2] Copyright 1999, with permission from Elsevier (the scale bar is 1 μm))

was obtained by this technique, shows fat droplets and casein micelles in the matrix.

Optical Microscopy

In optical microscopy the contrast between different components arises from differences in their refractive indices. Ice cream is too opaque for its structure to be observed. However, samples can be prepared so that the ice crystals, air bubbles and fat droplets can be visualized separately. Lactose crystals, if any are present, can be clearly observed using crossed polars. They have a characteristic arrowhead shape that makes them easily distinguishable.

Careful temperature control is necessary so that ice crystals neither melt nor grow while they are imaged. This can be achieved by performing the sample preparation and microscopy inside temperature-controlled chambers or by carrying out the whole experiment in a cold room. The temperature of the cold chamber is set to the required imaging temperature, for example −5 °C for ice cream that has just come out of the factory freezer, or −18 °C for ice cream that has been hardened and stored. A small sample of ice cream is smeared onto

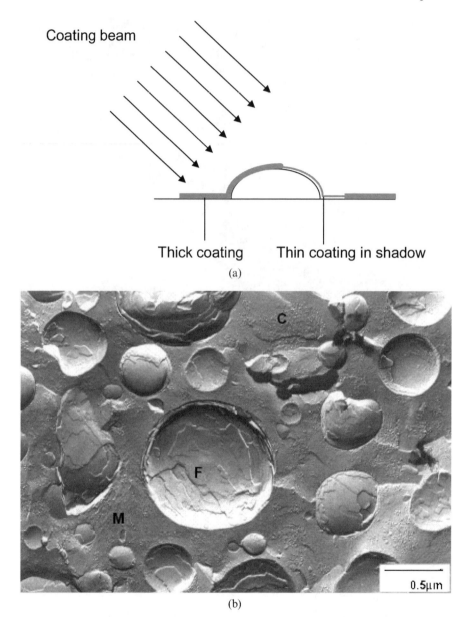

Coating beam

Thick coating   Thin coating in shadow

(a)

(b)

Figure 6.3   (a) *Schematic diagram of shadowing*; (b) *freeze–fracture transmission elec-
            tron micrograph of ice cream showing fat droplets* (F) *and casein micelles*
            (C) *in the matrix* (M)

a microscope slide, a drop of cold solvent, such as white spirit, butanol
or ethyl acetate, is added and a cover slip is placed on top and pressed
down. This process destroys most of the air cells and separates the ice
crystals so that they do not lie on top of each other. The refractive index

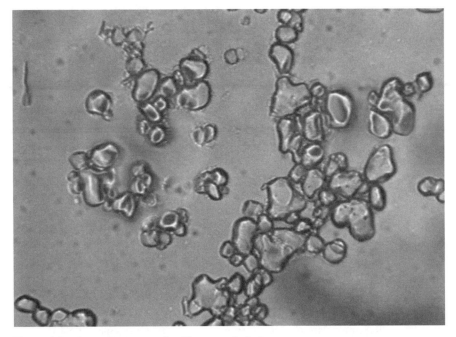

Figure 6.4 *Optical micrograph of ice crystals in ice cream*
(The image is 500 μm wide)

of the solvent is close to that of the fat and matrix, so the ice crystals
stand out. The slide is transferred onto a cold stage on the microscope,
also at the required temperature. The sample is imaged using conventional bright field light microscopy with a typical magnification of ×40.
If the magnification is too low, small ice crystals may be missed,
whereas if it is too high it may not be possible to capture large ones
within the field of view. Figure 6.4 shows an optical micrograph of ice
crystals obtained by this method.

The ice crystals are larger in water ices, and the ice content is higher
than in ice cream. This means that the crystals are harder to disperse,
so the solvent dispersion/squashing step may need to be repeated.
Polarized light microscopy can be used to distinguish overlapping ice
crystals.

To visualize air bubbles, a thin slice of ice cream is placed on a microscope slide at room temperature. A drop of viscous liquid, such as
glycerol, is added and a cover slip is placed on top. This melts the ice
crystals but prevents the air cells coalescing. The sample is imaged using
conventional bright field light microscopy at a typical magnification
of ×40. Figure 6.5 shows an optical micrograph of air bubbles obtained
by this method. The air bubbles are all spherical because they are

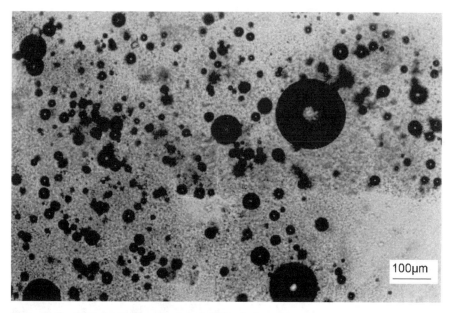

Figure 6.5 *Optical micrograph of air bubbles in ice cream*
(Reprinted with permission from *Microsc. Anal.* September 1994[3])

no longer surrounded by ice crystals and viscous matrix; this is not an accurate representation of their shape in ice cream. However, their size distribution is representative.

At about 1 μm, fat droplets are close to the resolution limit of optical microscopy, and therefore are harder to observe than ice crystals or air bubbles. They can be visualized by using a sophisticated imaging technique, such as differential interference contrast, together with high magnification. Samples are prepared by placing a drop of diluted mix or melted ice cream on a slide, and covering it with a cover slip. We have already seen images of fat droplets obtained by this technique in Figures 4.2 and 4.6.

The location of polysaccharides and proteins within the matrix can be visualized with optical microscopy by labelling them with a dye or a stain and using freeze-substitution (described in the section on TEM) to stabilize the structure. Polysaccharides can be labelled with a dye that fluoresces when bright light is shone on it. The light is viewed through a filter so that only the emitted wavelength is seen. Thus, areas of the sample that contain the dye-labelled component appear bright against a

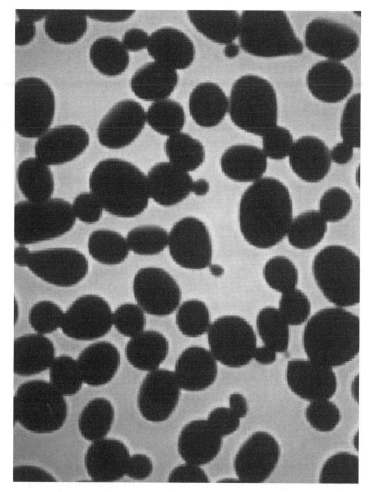

Figure 6.6 *Fluorescent microscopy image of an unaerated model ice cream*
(The image is 80 µm wide)
(Reprinted from 'Fluorescence Microscopy to Study Galactomannan
Structure in Frozen Sucrose and Milk Protein Solutions' Copyright 1999
with permission from Elsevier,[4])

dark background. Figure 6.6 shows an image of an unaerated ice cream
sample obtained by this technique. The fluorescent dye resides in the
matrix, which appears bright. The dye is excluded from the ice crystals,
which are black.

Confocal Laser Scanning Microscopy

Confocal laser scanning microscopy (CLSM) uses labelling to obtain
three-dimensional images. CLSM has not yet been used extensively on

ice cream, but has great potential. A laser beam is focussed on and scanned across a plane at a known depth in the sample. When the laser beam strikes the dye-labelled components they emit fluorescent light that is focused onto a photo-detector. The optics are arranged so that reflected light and fluorescent light from outside the focal plane is blocked. A 2D image is formed of a 0.1 μm thick slice of the sample in the focal plane, and a 3D image of the microstructure can be reconstructed on a computer by stacking the sequence of 2D images taken at different depths.

Measurement of Fat Structure

In addition to optical microscopy, there are other methods by which the fat structure can be studied. Laser light scattering provides a method for measuring the fat particle size distribution. Samples are diluted in distilled water so that fat droplets are well separated in solution. If required, a small amount of a surfactant can be added to break up flocculated aggregates of fat droplets. This facilitates analysis of the scattering pattern, from which the particle size distribution is calculated.

Turbidity and solvent extraction are relatively quick and easy measurement techniques that give less detailed information than microscopy and light scattering. They are used to measure the amount of destabilized fat (DSF), rather than the particle size distribution. Turbidity measures the reduction in intensity (the absorbance) of a beam of light when it is shone through a dilute solution of melted ice cream. For a fixed total amount of fat, large fat clusters have a smaller total surface area than small droplets. Therefore, they scatter less light, so the absorbance is lower. Comparing the absorbance of samples of mix (a_{mix}) and melted ice cream (a_{melt}) gives a measure of the amount of destabilized fat.

$$\text{DSF}(\%) = \frac{a_{mix} - a_{melt}}{a_{mix}} \times 100 \qquad (6.1)$$

The second method uses a solvent to extract the de-emulsified fat preferentially: the less emulsified the fat, the easier it is to extract. Approximately 10 g of mix or melted ice cream and 30 ml of a solvent, such as heptane, are placed in a flask. The flask is rapidly rotated to dissolve the de-emulsified fat in the solvent and then left to stand so that the solvent separates out. The solution of the fat in the solvent is then decanted, and the solvent is removed by drying. The mass of the remainder, *i.e.* the extractable fat, is determined ($m_{\text{extracted}}$) and expressed as a percentage of the total mass of fat (m_{total}).

$$\text{DSF}(\%) = \frac{m_{\text{extracted}}}{m_{\text{total}}} \times 100 \tag{6.2}$$

Image Analysis and Microstructural Characterization

Once samples have been visualized by any of the microscopy techniques, images are recorded, most conveniently in digital form on a computer. It may be sufficient simply to look at the image, but if more detailed information is required, such as the ice crystal size distribution, it is necessary to analyse the image and characterize the microstructure. The first step is to segment the image, that is to identify and distinguish the microstructural components. This can be done manually, by drawing around the component of interest, or automatically by computer, provided that there is sufficient contrast between the phases. Figure 6.7 shows a SEM micrograph of a slush-frozen water ice that has been segmented into its components of ice and matrix. The white lines mark the boundaries between the components.

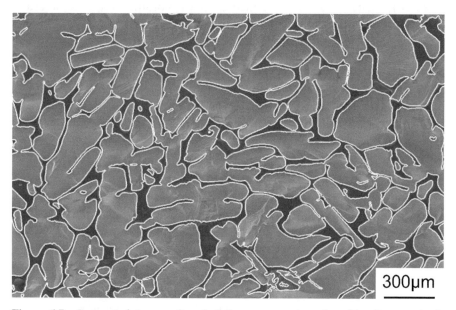

Figure 6.7 *Segmented image of a slush-frozen water ice; the white lines mark the boundaries between the ice crystals and matrix*

Once an image has been segmented, the microstructural characteristics can be measured, for example ice crystal and air bubbles size. Figure 6.1 shows that the ice crystals and air bubbles are not spherical. Therefore, several different measurements of size can be made, for example the maximum diameter, the minimum diameter, or the equivalent circular diameter (*i.e.* the diameter of a circle with the same area). The shape factor (defined by equation 6.3) is a measure of how far a particle is from being circular: it takes a value of one for a sphere and becomes smaller as the particle becomes less spherical.

$$\text{shape factor} = \frac{4\pi \times \text{area}}{(\text{perimeter})^2} \tag{6.3}$$

The ice crystals and air bubbles are not all the same size, so it is necessary to determine their distribution, that is the proportions of particles of different sizes. To obtain a statistically accurate size distribution, many particles (typically a few hundred) must be measured. This may require several images to be obtained for each sample. It is generally more convenient to quote a mean size rather than the whole distribution. There are several different ways of calculating the mean. The simplest is the number average size, that is the sum of all the sizes of the particles, divided by their total number. However, a few very small particles can skew the distribution so that the number average diameter gives a misleading result. Surface area or volume weighted average diameters give a more representative average in this case.

As well as the size and shape of the microstructural components, their connectivity has a large effect on the physical and sensory properties. One measure of the connectivity is the contiguity. Figure 6.8 shows a schematic diagram of a water ice microstructure (*i.e.* with two phases, ice and matrix). A line is drawn across the image and the numbers of ice–ice (n_{ii}) and ice–matrix (n_{im}) contacts are counted per unit length. The contiguity of the ice (C_{ii}) is given by

$$C_{ii} = \frac{2n_{ii}}{2n_{ii} + n_{im}} \tag{6.4}$$

Since the contiguity depends on where the line is drawn, it is usual to draw several lines and calculate the mean. Some examples of the effects of the connectivity are discussed in Chapter 7.

Computer Simulations of Microstructure

The process of making samples, visualizing the microstructure and measuring the physical and sensory properties is expensive and

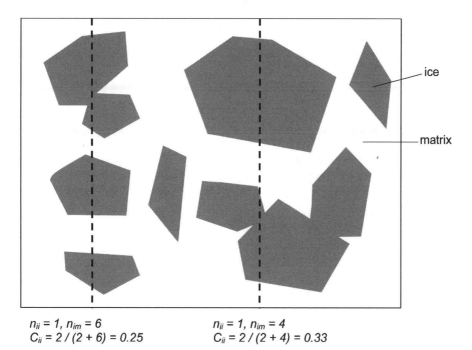

$n_{ii} = 1, n_{im} = 6$
$C_{ii} = 2 / (2 + 6) = 0.25$

$n_{ii} = 1, n_{im} = 4$
$C_{ii} = 2 / (2 + 4) = 0.33$

Figure 6.8 *Schematic water ice microstructure showing how the contiguity is calculated*

time-consuming. In recent years it has become possible to use computer simulations as an alternative or a complement to experimentation in many areas of science, including modelling the changes in the microstructure of water ices during hardening. Figure 6.9 shows simulated microstructures of a slush-frozen water ice. The light regions are ice

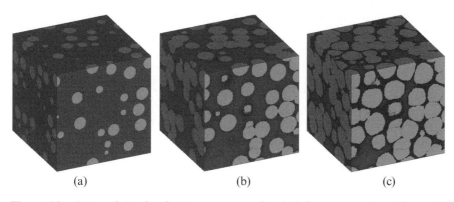

(a) (b) (c)

Figure 6.9 *Series of simulated microstructures of a slush-frozen water ice:* (a) *on exit from the factory freezer;* (b) *part way through hardening; and* (c) *at the end of hardening*
(Courtesy of Dr J. Aldazabal, CEIT, Spain)

crystals and the dark background is the matrix. In the first image there are a number of small, separate ice crystals, which grow and come into contact with others as the temperature is lowered during hardening and the ice content increases. Mechanical and thermal properties can be calculated from the simulated microstructures.

MECHANICAL AND RHEOLOGICAL PROPERTIES

The mechanical and rheological properties are measures of how materials respond when they are deformed. Solids are characterized by mechanical properties such as the Young's modulus, strength and hardness, whereas liquids are characterized by rheological properties such as the viscosity and yield stress.

When a stress (*i.e.* a force per unit area) is applied to a solid, for example it is stretched, sheared, twisted or squashed, it deforms, *i.e.* changes its length, or shape. For small deformations, the amount of deformation is proportional to the applied stress. The material is said to behave elastically. Beyond a certain deformation (the elastic limit) the material ceases to be elastic, and the material no longer returns to its initial shape when the stress is removed. This is called plastic deformation. If the material is deformed further then it will eventually break. Some materials, such as rubber, are elastic for large deformations, while others, such as plasticine, have a relatively small elastic limit but can then undergo large plastic deformations. Brittle materials, such as china, can only withstand small deformations before they break.

Figure 6.10 shows two ways of deforming a solid. Figure 6.10a shows deformation by stretching. In the elastic region, the fractional change in length (the strain $\Delta l/l$) is proportional to the stress, and the constant of proportionality is the Young's modulus, Y.

$$\sigma = Y \frac{\Delta l}{l} \tag{6.5}$$

A shear deformation changes the shape of the material (Figure 6.10b). The shear strain, *i.e.* the displacement in the direction of the force divided by the height of the block, $\Delta l/h$, is proportional to the stress, and the constant of proportionality is the shear modulus, G.

$$\sigma = G \frac{\Delta l}{h} \tag{6.6}$$

Liquids have no definite shape, and flow irreversibly when a shear stress is applied. We have already seen the definition of the viscosity of a liquid in Chapter 2 in terms of the shear stress and velocity gradient.

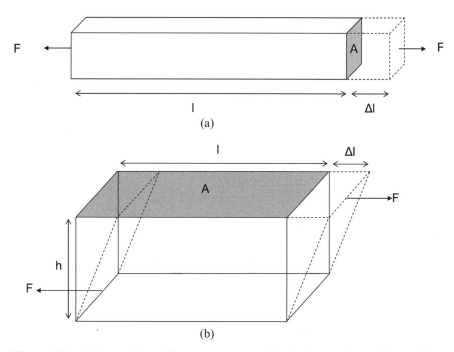

Figure 6.10 (a) *A stretching deformation, associated with the Young's modulus and* (b) *a shear deformation, associated with the shear modulus*

Since the velocity (*v*) is the distance travelled (Δ*l*) divided by the time (*t*), equation 2.10 can be rewritten as

$$\sigma = \eta \left(\frac{\Delta l}{h} \right) \Big/ t \tag{6.7}$$

(Δ*l*/*h*)/*t* is the rate of change of the shear strain, *i.e.* the shear rate. Comparing equations 6.6 and 6.7, we can see that there is a correspondence between the shear modulus and shear strain on one hand, and the viscosity and shear rate on the other. Stressing a solid produces a certain shear strain, whereas stressing a liquid causes it to flow with a certain shear rate. Materials, such as ice cream, which display properties of both solids and liquids, are described as viscoelastic. To describe their response to deformation, two different sorts of measurements, are required: mechanical properties for the solid-like characteristics, and rheological measurements for liquid-like ones.

Mechanical Properties

At a typical storage temperature of $-18\,^{\circ}\text{C}$, ice cream displays solid-like properties such as elasticity, plasticity and fracture. A number of different tests that deform a sample in different ways are used to measure these, such as bend testing, compression testing and hardness testing. They are performed using instruments that deform the sample in a controlled manner while recording the applied force and the resulting deformation, for example an Instron[TM] or a texture analyser. Since the mechanical properties of ice cream are very sensitive to temperature, a temperature-controlled sample environment is required.

Three-point Bend Test

Bend testing is a standard method in materials science for measuring mechanical properties such as the Young's modulus and the strength, which can be applied to ice cream. A bar of ice cream or water ice is prepared in a mould of fixed dimensions, typically 25 mm depth (D) × 25 mm breadth (B) × 200 mm span (S). Three metal rods hold the bar as shown in Figure 6.11. The top two are fixed, while the bottom one is slowly moved upwards, bending the bar. The force (F) required is plotted against the resulting displacement (d) until the bar breaks.

Figure 6.12 shows a typical force–displacement curve for ice cream. The force initially increases linearly with the displacement (this is the elastic region). The Young's modulus is a measure of the resistance to deformation of the ice cream and is given by the initial slope of the force displacement curve, $\Delta F/\Delta d$. At a certain point, the slope of the force–displacement curve begins to decrease. This corresponds to the onset of plastic deformation. The maximum force (F_{max}) corresponds to the point at which a crack forms and the sample begins

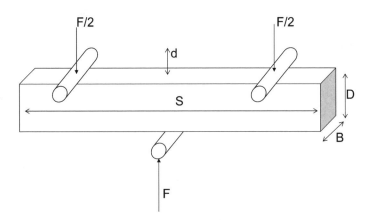

Figure 6.11 *The three-point bend test*

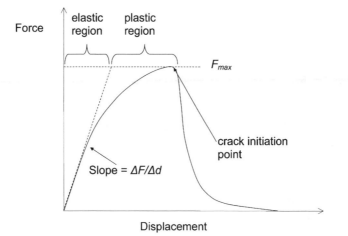

Figure 6.12 *Force–displacement curve from the three-point bend test*

to fail, and is a measure of the strength of the ice cream. When the sample breaks, the force returns to zero. The decrease of the force after reaching the maximum may be very fast (for a brittle sample) or gradual (for a plastic sample).

Equations 6.8 and 6.9 give the Young's modulus and the strength in terms of the parameters obtained from the force–displacement curve and the dimensions of the bar.

$$\text{Young's modulus} = \left(\frac{\Delta F}{\Delta d}\right)\frac{S^3}{4BD^3} \tag{6.8}$$

$$\text{Strength} = \frac{3F_{\text{max}}S}{2BD^2} \tag{6.9}$$

Compression Test

The compression test is another means of measuring the mechanical properties of ice cream. Cylinders, with an initial height, h_0, (typically 4 cm), and cross-sectional area, A_0 (typically 10 cm²), are placed between two plates (Figure 6.13a) and compressed by moving one plate towards the other (Figure 6.13b) at a constant speed (typically 50 mm min⁻¹). This is similar to the deformation that takes place in the mouth during eating, for example when ice cream is squashed between the tongue and the roof of the mouth. Friction between the plates and the sample would cause it to become barrel shaped when it is compressed, so the plates are lubricated to ensure that the sample remains

(a)

(b)

Figure 6.13 *Ice cream sample at* (a) *the start and* (b) *the end of a compression test*

approximately cylindrical. During compression the cross-sectional area (A) increases and the height (h) decreases, but the volume of the sample (V) is constant, *i.e.*

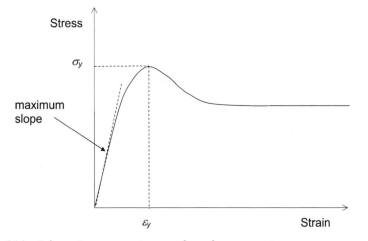

Figure 6.14 *Schematic stress-strain curve from the compression test*

$$V = h_0 A_0 = hA \tag{6.10}$$

The compression force (F) is recorded as a function of the height, from which the stress (σ) and strain (ε) are calculated.

$$\sigma = \frac{F}{A} = \frac{F}{A_0} \times \frac{h}{h_0} \tag{6.11}$$

$$\varepsilon = -\ln\left(\frac{h}{h_0}\right) \tag{6.12}$$

Figure 6.14 shows a stress–strain curve from which the mechanical properties are obtained. The maximum slope gives the compressive Young's modulus, the maximum stress is the yield stress (σ_y) and the corresponding strain is the yield strain (ε_y). A high yield stress indicates a strong material.

Hardness Test

Hardness is a measure of the ability of a material to resist plastic deformation. In a hardness test, an indenter is driven into the material. The force is recorded as a function of the depth of penetration. Unlike, for example the Young's modulus, hardness is the result of a particular measurement procedure rather than an intrinsic property of the material. A typical measurement consists of inserting a cylindrical probe a few millimetres in diameter into a block of ice cream at $-10\,^{\circ}\text{C}$

to a depth of about 1 cm. This requires a force of a few Newtons. The hardness (H) is given by

$$H = \frac{F_{max}}{A} \qquad (6.13)$$

where F_{max} is the maximum force and A is the area of indentation. The area depends on the shape of the indenter, and for some indenter shapes, such as a pyramid or a sphere, it also depends on the indentation depth. Experiment 15 in Chapter 8 describes a simple method for comparing the hardness of ice cream samples.

Rheological Properties

Rheometers are used to measure rheological properties of liquids. They apply a controlled shear stress or strain to the sample and measure the response, from which the rheological properties, such as apparent viscosity or yield stress, are calculated. Rheometers consist of four basic elements: (i) a sample container with a certain geometry; (ii) a device that applies a controlled shear stress or shear rate (either continuous or oscillating) over a wide range of values; (iii) a device to determine the response; and (iv) a means for controlling the temperature. It is important to be able to measure the rheology across the whole range of temperatures and shear rates that ice cream experiences during production and consumption.

The most commonly used geometries are shown in Figure 6.15. Concentric cylinders are suitable for mix or melted ice cream; one cylinder is rotated and the torque on the other is measured. Parallel plates are used for ice cream at relatively high temperatures. Discs (typically

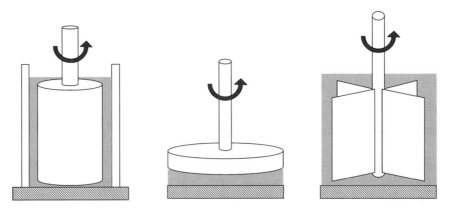

Figure 6.15 *Concentric cylinder, parallel plate and vane geometries*

40 mm diameter and 5 mm thick) are cut from a block, placed between the plates and allowed to equilibrate at the measurement temperature. To make the measurement one of the plates is rotated and the torque on the other is measured. However, the parallel plate geometry is not suitable for ice cream at low temperatures. The ice cream would be too viscous for reliable measurements and would slip at the surface of the plate. The vane is therefore used at low temperatures. It is inserted into a cup of ice cream and torque is applied. Above a certain torque the vane begins to rotate when the ice cream yields and starts to flow. The yield stress is calculated from the maximum torque applied to the vane and its dimensions.

In continuous tests the deformation is large so the structure of the sample is destroyed. In oscillatory rheometry the deformation is small so the structure remains intact. Both viscous and elastic properties can be measured simultaneously. The elastic, solid-like component of the response (the storage modulus, G') is in phase with the deformation and the viscous, liquid-like component (the loss modulus G'') is out of phase. The ratio G''/G' is a measure of the relative importance of the viscous and elastic components. Thus, for example, ice cream that has a high G' and a low G''/G' is more solid-like than liquid-like.

THERMAL PROPERTIES

The thermal properties of ice cream, such as the heat capacity and the thermal conductivity, are important for several reasons. Parameters of the production process, such as the length of time required to harden the ice cream, depend on the thermal properties, as does the rate at which ice cream warms up and melts. This is important in storage and distribution and also when the ice cream is consumed. It should not melt so rapidly that it falls off the stick before it can be eaten. The thermal properties also affect the sensory properties such as the perception of coldness in the mouth. Furthermore, the freezing point and glass transition lines on the phase diagram, and the ice curve, are obtained from measurements of the thermal properties. The main techniques for measuring the thermal properties are calorimetry, conductivity measurements, thermal mechanical analysis and meltdown.

Calorimetry

The heat capacity, freezing point curve and the ice curve of ice cream can be determined by calorimetry. In adiabatic calorimetry, the sample is held in an insulated chamber. A controlled amount of heat is input and the resulting increase in temperature is measured, from which the heat capacity, freezing point and ice content are calculated. Figure 6.16a shows the heat capacity of a sample of ice cream as a function of temperature, and Figure 6.16b shows the ice curve. Below $-15\,^{\circ}\mathrm{C}$ the heat capacity is nearly constant. Between -15 and $-2\,^{\circ}\mathrm{C}$ the ice content decreases rapidly. This appears as a large peak in the heat capacity. Instead of increasing the temperature, the input heat is absorbed as latent heat by the melting ice. At $-2\,^{\circ}\mathrm{C}$, the last ice melts. The heat then causes the temperature of the melted ice cream to increase. The heat capacity of melted ice cream is larger than that of frozen ice cream because the heat capacity of water is larger than that of ice.

Thermal Conductivity

The basic method of determining the thermal conductivity is to place a sample of known dimensions in a temperature gradient and measure the rate of the resulting heat flow through it. Suitable apparatus consists of a disc of ice cream sandwiched between two plates made from a material of known thermal conductivity and placed in an insulated cylinder. One plate is heated and the other cooled to produce the temperature gradient, which is measured with thermocouples embedded in the plates and sample.

Thermal Mechanical Analysis

The glass transition temperature of ice cream can be measured by thermal mechanical analysis. A sample of unaerated frozen ice cream mix approximately 0.5 mm thick and 1 cm in diameter is placed between two plates. One of the plates is attached to a probe that measures the expansion of the sample as it is warmed up from below the glass transition temperature. Unaerated samples are used because aeration affects the thermal expansion. However, since the glass transition temperature is a property of the composition of the matrix, the absence of air does not affect the result. The glass transition temperature is indicated by a change in the rate of expansion that occurs at about $-30\,^{\circ}\mathrm{C}$ for a typical ice cream.

Meltdown

The meltdown test (or melt resistance) of ice cream measures its ability to resist melting when exposed to warm temperatures for a

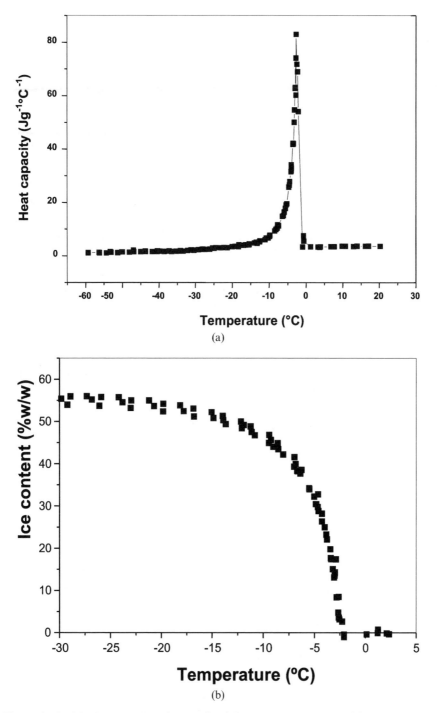

Figure 6.16 (a) *Heat capacity of a sample of ice cream as it is warmed from −60 °C to 20 °C in an adiabatic calorimeter and* (b) *the corresponding ice content*

period of time. Unlike the other thermal properties, meltdown is not a uniquely defined standard physical parameter. It is actually an empirical measure that reflects a number of factors, including thermal conductivity, heat capacity, microstructure and formulation. It can indicate, for example, the effect of changing the formulation or process on the properties of the ice cream.

A standard block of ice cream (*e.g.* 1 l) initially at the storage temperature (*e.g.* −18 °C) is placed on a wire mesh in an enclosed chamber that is held at a warm temperature, *e.g.* 25 °C, and allowed to melt. A funnel is placed beneath the mesh to collect the melted ice cream, which drains into a graduated cylinder, or a beaker on a balance. The volume or mass of melted ice cream is recorded as a function of time during melting and is plotted as a percentage of the original volume or mass. Figure 6.17 shows a series of images of a block of ice cream during a meltdown test, together with the corresponding meltdown curve. The meltdown test is used both as a research and development tool and also as a quality control measure.

SENSORY PROPERTIES

The sensory properties are the characteristics of foods perceived by the senses of sight, smell, taste, touch and hearing, such as flavour, texture and appearance. The human sensory organs are a remarkably sensitive means of measuring sensory properties.

Flavour is a complex combination of aromas, smelt through the nose (*i.e.* volatile compounds), and taste, experienced by the tongue (*i.e.* non-volatile liquids or solids). Aromas are perceived *via* the olfactory nerve endings in the upper part of the nose, often in very small concentrations (a few milligrams per kilogram of product). The importance of aromas to a flavour can be demonstrated by pinching your nose as you eat: the food will taste blander. The number of volatile compounds responsible for an individual aroma can be very high: there are between 100 and 500 compounds in aromas such as vanilla, strawberry and cocoa. Most adults can detect the difference between about 2000 aromas. Typically taste compounds must be present in concentrations as much as 1000 times higher than aromas. They can only be detected by the taste receptors on the tongue when they are in solution, so dry substances must first be dissolved in saliva. Taste buds are sensitive to five basic stimuli: bitter (*e.g.* caffeine and quinine); sour (*e.g.* lemon juice); salt; sweet (*e.g.* sugar) and umami (*e.g.* monosodium glutamate). Interestingly, stereoisomers of chiral molecules can have completely different

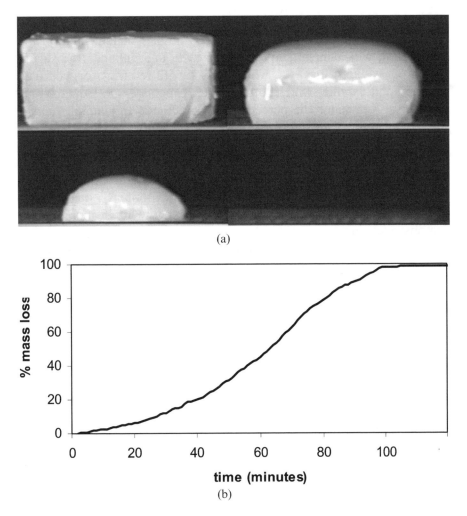

(a)

(b)

Figure 6.17 (a) *Series of images of a block of ice cream after 0, 30, 70 and 100 min during a meltdown test and (b) the corresponding meltdown curve*

flavours. For example, one stereoisomer of carvone has the taste of caraway and the other spearmint. Similarly, the two forms of limonene give the taste of orange and lemon. Chirality also affects the aroma; stereoisomers usually have different smells, and one form often has no smell at all.

In addition to the compounds that provide aroma and taste, two other types of molecule can affect flavour: molecules that affect the trigeminal (touch) receptors, and flavour enhancers. Trigeminal sensations in the nose, lips, mouth and throat are responsible for the coolness of menthol, the 'bite' of mustard and pepper, and the warmth of cloves.

Flavour enhancers and suppressers are used in low concentrations to enhance or suppress other flavours. Examples include maltol and ethylmaltol, which have a low caramel taste and enhance the sugary feeling of products; furaneol, which is used with red fruits or wild fruit flavours; and vanillin, which softens bitter chocolate and fruit flavours and can also enhance the perception of sweetness. In general, sucrose suppresses bitter, sour and salty tastes, for example in chocolate, and enhances fruit flavours. A further important point for ice cream is that the perception of flavour is affected by temperature: flavours are less intense at low temperatures. For this reason, ice cream and water ices are generally more strongly flavoured than products consumed at warmer temperatures, such as soft drinks (Experiment 17 in Chapter 8 demonstrates this).

The sensory perception of texture is produced by a combination of the neural signals from pressure sensors in the tongue and mouth and from the sensors in the muscles involved in mastication that sense their forces and displacements. The perception of texture arises not only from the properties of the original sample, but also from how the sample breaks down in the mouth as it is eaten.

Two types of sensory assessment are used for ice cream, analytical methods, in which trained tasters provide quantitative sensory data, and consumer testing, in which consumers are asked questions such as how much they like a product and whether they would be likely to buy it.

Analytical Methods

Analytical methods can be sub-divided into difference tests and descriptive analysis. Difference tests such as triangle and paired comparison tests are designed to identify differences between samples. In a triangle test, the assessor is given three samples, two of which are the same, and asked to identify which sample is different. In a paired comparison test, the assessor is asked to identify whether there is a difference in a particular sensory characteristic between a pair of samples. Descriptive analysis, in which the sensory characteristics of a sample are described and scored on a scale, is probably the most important analytical method.

Quantitative descriptive analysis (QDA®) is a method that is often used. A panel of tasters first assesses a set of training samples and scores the intensity of a number of sensory characteristics (attributes) of the samples. There are many different attributes, and there is no unique, exhaustive list, but some that are commonly used to describe the texture of ice cream are listed below. Once trained, the panellists score the actual samples for each attribute. Statistical methods are then used to identify whether differences in the scores of an attribute for different samples are significant.

A simple vanilla ice cream is often used when evaluating texture in order not to complicate the assessment with other flavours. If the appearance of the samples is different, for example one is white whereas another is yellow, then the lighting can be arranged so that the samples appear similar. This prevents the appearance unconsciously affecting the perception of texture. A series of samples should be presented in balanced random order so that the panellists do not know which to expect. The ice cream is usually expectorated after each sample, and the mouth rinsed with room temperature water to cleanse the palate and to prevent it from becoming too cold. Each sample is usually evaluated several times by all panellists to check that scoring is consistent.

Attributes are associated with different stages of consumption so a standard procedure for eating the ice cream is specified, together with which attributes are assessed at each stage. Ice cream is initially spooned, then bitten, then squashed with the tongue and moved around the mouth, and finally swallowed. It is difficult to judge a number of different sensory attributes simultaneously, so several small samples are usually taken in turn. It is also possible to measure how the intensity of a single attribute varies with time during eating. In this case, the panellist's score is recorded as a function of time, for example on a computer by moving a pointer along a line that represents the intensity of the attribute.

Attributes perceived before eating:
- *Firmness* (or *Hardness*) *on spooning* is the resistance felt when pushing a spoon into the sample
- *Crumbliness* is a measure of how readily the ice cream breaks apart, for example on spooning.

Attributes perceived in the initial stages of eating:
- *Firmness in the mouth* is the resistance felt when squashing the sample with the tongue.
- *Chewiness* or *Gumminess* is the degree to which the sample is chewy or elastic on first chewing. Excessive chewiness may be the result of too much stabilizer. *Weakness*, a lack of resistance to manipulation in the mouth is the opposite of chewiness.
- *Lightness* and its converse *Density* are related to the amount of air.

- *Coarseness* or *Iciness* is a lack of *Smoothness*, usually due to presence of large ice crystals.
- *Coldness* is the degree of cold sensation in the mouth. It is related to the amount of heat absorbed from the ice cream by the mouth, and hence to the ice content and thermal properties.
- *Rate of melt* is the speed at which the sample can be reduced to a liquid. This is perceived in the mouth, rather than physically measured in the meltdown test. Fast melt is often associated with weakness.

Attributes perceived in the later stages of eating:
- *Thickness* is related to the viscosity of melting ice cream before swallowing.
- *Mouth-coating* is a measure of how the melting ice cream coats the inside of the mouth on or just after swallowing.

Attributes associated with particular defects:
- *Sandiness* refers to the perception of lactose crystals.
- *Waxy mouth-feel* is associated with the presence of high melting point fats that do not quickly melt in the mouth.

Some terms that are commonly used by consumers are not suitable as sensory attributes since they are really combinations of a number of different attributes. For example, creaminess is a combination of smoothness, thickness and creamy flavour. Experiment 16 in Chapter 8 gives some suggestions for comparing the sensory attributes of ice cream samples.

QDA describes a number of different sensory characteristics, some of which are likely to be related to others, for example, coarseness and iciness are opposed to smoothness, weakness is related to the rate of melt, and thickness is related to mouth-coating. Therefore, while an experiment might assess ten different sensory characteristics, some of these will be related, so there are not actually ten independent variables. Principal component analysis (PCA) is a statistical method by which a large number of variables can be reduced to two or three truly independent principal components (PCs). PCA (and other statistical methods such as regression analysis and significance testing) can be used for both sensory and physical characteristics, and to look for correlations between them. The variables are plotted on a diagram that has the principal components as axes. (When the variables are sensory attributes this is sometimes called 'sensory space'.) Samples are represented by points on the PCA diagram.

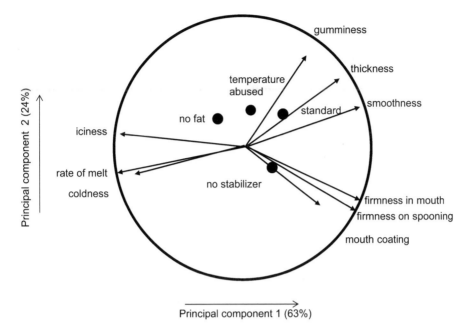

Figure 6.18 *Principal component analysis diagram for four different ice cream samples*

PCA allows the visualization of the differences between samples in relation to the sensory attributes. Figure 6.18 shows a PCA diagram for a sensory assessment of four ice cream samples: fresh ice cream, ice cream that has been subject to temperature abuse, zero fat ice cream and ice cream without any stabilizer. The arrows show the directions corresponding to the sensory attributes relative to the two principal components, PC1 and PC2. Arrows that do not reach the edge of the circle indicate that the attribute also has a contribution from PC3, which is not shown. However the contribution of PC3 is small: PC1 and PC2 together account for 87% of the variance of the data in this case. Coldness and iciness are very close to the negative PC1 direction; smoothness very close to the positive PC1 direction. Thus they are opposite ends of the same axis: an ice cream that is icy is not, and *vice versa*. PC1 is clearly related to the ice content and ice microstructure. Other attributes, for example, gumminess, thickness and mouth-coating lie between the *x* and *y* axes, *i.e.* they are combinations of PC1 and PC2. From the locations of the four samples on the diagram it is apparent that low fat content and temperature abuse both make ice cream icier and colder. Removing the stabilizer makes the sample less thick and gummy, and more mouth-coating. This suggests that PC2 for this set of samples is strongly influenced by the stabilizer. All the samples have 100% overrun and as a result there is little difference between the samples in terms of attributes such as lightness that relate to the air phase volume. Therefore, these attributes do not appear in the first and second principal components. Samples with different formulations or overruns would produce a different PCA diagram, on which the attributes would correspond to different directions.

In addition to flavour and texture measurements, the appearance of ice cream can be assessed, for example, with respect to the colour, and evenness, either by eye, or by using a camera in controlled lighting followed by image analysis.

Consumer Methods

It is unlikely that a group of untrained consumers will all assess samples and attributes in the same manner, so it is not possible to obtain reliable quantitative information from them. However, consumers are very good at saying whether they like a product, and whether they would buy it. It is also important to understand how sensory attributes drive consumer preference: for example how soft or sweet most people like their ice cream to be. Several different methods are used to assess these questions. These include focus groups, in which consumers' opinions on products are sought, preference ranking, where consumers indicate which of a number of products they like most, hedonic scaling, where they score, for example on a scale of 1 to 10, how much they like each sample, and purchase intent scaling, where they score how likely they would be to purchase a product at a particular price.

REFERENCES

1. C.J. Clarke, *Phys. Educ.*, 2003, **38**, 248.
2. H.D. Goff, E. Verespej and A.K. Smith, *Int. Dairy J.*, 1999, **9**, 817.
3. A. Sztehlo, *Microsc. Anal.*, September 1994, 7.
4. H.D. Goff, D. Ferdinando and C. Schorsch, *Food Hydrocolloids*, 1999, **13**, 353.

FURTHER READING

K.B. Caldwell, H.D. Goff and D.W. Stanley, *Food Struct.*, 1992, **11**, 1.
J.F. VelezRuiz and G.V.B. Canovas, *Crit. Rev. Food Sci. Nutr.*, 1997, **37**, 311.

Chapter 7

Ice Cream: A Complex Composite Material

Composites are materials made of a combination of two or more substances on a microscopic scale. A good example of a composite material comes from the aerospace industry. Bearings for jet engines have to be strong and tough at temperatures greater than 1000 °C. At these temperatures most metals weaken or melt. Ceramics maintain strength at high temperatures but are brittle. However, a composite material can combine the strength of ceramics at high temperatures with the fracture resistance of metals. The properties of a composite are determined both by the properties of the individual components and by the way that they are combined, *i.e.* the microstructure. A bearing made from a single piece of metal joined to a single piece of ceramic would suffer from the shortcomings of both and have the strengths of neither. However, when the materials are mixed at a microscopic scale the desired properties are obtained.

Ice cream is also a composite material. The main ingredients provide the required sensory properties: ice gives cooling, fat provides creaminess, air gives lightness and softness, sugar provides sweetness, and flavours enhance its taste. However, if you simply put ice cubes, whipped cream, sugar and vanilla essence in a bowl and stir you end up with a mess that bears no resemblance to ice cream. Even though you have used the correct ingredients, you do not produce the combination of cooling, creaminess, softness, sweetness and flavour in a single substance. This is because you have not created a composite material with the right microstructure of ice crystals, air bubbles, fat droplets and matrix that we saw in Figure 6.1.

Figure 7.1 shows how the microstructural components of ice cream relate to the ingredients. The air bubbles are made up of air together

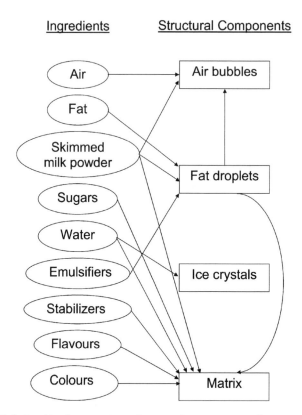

Figure 7.1   *Relationships between ingredients and microstructural components*

with milk proteins and fat droplets that coat the surface. The fat droplets in turn are made up of the fat, milk protein and emulsifiers. Ice crystals are simply formed from water, and the matrix is a solution of sugars, milk proteins, stabilizers, flavours and colours, in which some of the fat is suspended.

   Both the total amount of each component and the microstructure (*i.e.* the sizes, shapes and connectivity of the particles) are important. Together they determine the properties of the composite, *i.e.* the physical and sensory properties of ice cream. The amounts of the structural components are different for different types of ice cream. Table 7.1 shows typical volume fractions of each component at −18 °C in standard, premium, low fat and soft scoop ice cream, and water ice. Premium ice cream contains more fat than a standard ice cream, whereas a low fat ice cream (obviously) contains less. Soft scoop ice cream contains less ice than the standard. Water ice is a composite of two materials, ice crystals and matrix, and contains no air or fat.

Table 7.1 *Typical volume fractions of the structural components for different types of product*

Microstructural component	Standard ice cream	Premium ice cream	Low Fat ice cream	Soft Scoop ice cream	Water ice
Ice	0.30	0.35	0.31	0.23	0.75
Air	0.50	0.35	0.48	0.52	0
Fat	0.05	0.10	0.01	0.04	0
Matrix	0.15	0.20	0.20	0.21	0.25

Table 7.2 *Typical parameters of the structural components of ice cream*

Microstructural component	Volume fraction (%)	Size (µm)	Number (per litre)	Surface area ($m^2\ l^{-1}$)
Ice	30	50	5×10^9	40
Air	50	60	4×10^9	50
Fat	5	1	1×10^{14}	300
Matrix	15	–	–	–

Table 7.2 shows the typical volume fraction, particle size, number and surface area of each component in a standard ice cream.

In this chapter we look at each component in turn. However, many properties cannot be explained by simply looking at the separate components, so we next consider their interactions in the composite material. The microstructure of ice cream changes as it warms up and is manipulated by the mouth during consumption. Therefore, to understand the sensory properties, we also have to know how the microstructure breaks down during eating.

THE FOUR COMPONENTS

Ice

We have already seen that the amount of ice is determined by the formulation, in particular the sugars which produce the largest depression of the freezing point. Alcohol is also an effective freezing point depressant. If an alcoholic drink is used in ice cream (*e.g.* rum and raisin flavour) the amount of sugar must be reduced so that the ice content is not lowered. This ensures that the ice cream is not too soft.

The ice crystal size distribution as well as the ice content affects the properties of ice cream. Figure 7.2 shows the ice crystal size distribution

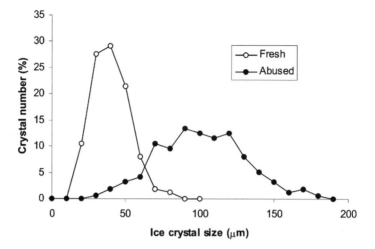

Figure 7.2   *Ice crystal size distribution in a sample of ice cream before and after thermal abuse*

in a sample of ice cream immediately after hardening and after delibe-rate temperature abuse by cycling between -20 and $-10\,^{\circ}\mathrm{C}$ every twelve hours for three weeks. The mean ice crystal size in the fresh sample is 40 µm. After temperature abuse this has increased to 100 µm and the distribution has become broader due to coarsening. We saw in Chapter 2 that recrystallization of a dispersion of ice crystals takes place by two mechanisms: accretion and Ostwald ripening. Recrystalli-zation results in an increase in the mean size and a decrease in the total number of ice crystals, while the overall ice phase volume is unchanged. The rate of recrystallization depends on the temperature: the lower the temperature, the slower the recrystallization.

We saw in Chapter 2 that recrystallization arises from the fact that small crystals have a larger surface area to volume ratio than large crys-tals, and hence are less stable. Another way of expressing this is to say that the solubility of a particle, $s(r)$, increases as its radius, r, decreases

$$\ln\left[\frac{s(r)}{s(\infty)}\right] = \frac{2\gamma V}{RT}\cdot\frac{1}{r} \qquad (7.1)$$

Thus, small crystals are more likely to dissolve or melt than large ones. Equation 7.1 is the Gibbs–Thomson equation. V is the molar volume and $s(\infty)$ is the solubility of an infinitely large particle.

Ostwald ripening can take place when the temperature is constant, but it is faster if the temperature fluctuates. When you take ice cream

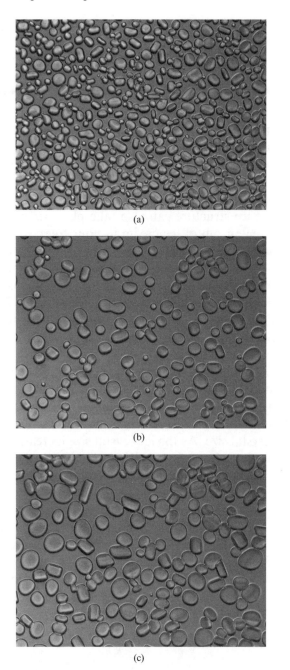

(a)

(b)

(c)

Figure 7.3 *Optical micrographs showing Ostwald ripening of ice crystals in a sucrose solution during temperature fluctuations:* (a) *Ice crystals at* $-10\ °C;$ (b) *warmed up to* $-7\ °C$ – *the smallest crystals melt;* (c) *then cooled back down to* $-10\ °C$ – *the ice reforms on the surviving crystals, which become larger* (The images are 240 μm wide)

home from the supermarket freezer its temperature may rise by a few degrees during the journey, so that some of the ice melts. Ice melts from all of the crystals, so crystals that are initially large become smaller, and those that are initially small disappear completely. When you put the ice cream in your freezer at home, it is cooled down again, and the ice that had melted freezes again. However, it cannot form on the crystals that have disappeared, nor is there sufficient supercooling to cause nucleation of new crystals. Instead it freezes on the crystals that have survived. The net effect is that the total number of crystals is reduced, and their mean size increases, while the total amount of ice is unchanged. This is shown in Figure 7.3.

Changes in the ice structure can also take place *via* water vapour. If you leave a half-eaten tub of ice cream in your freezer for a long time, you will find that a layer of frost will form on the inside of the tub. This is because ice sublimes from the ice crystals in the ice cream into the space in the tub, and then freezes again as frost on its inside surfaces.

The second coarsening mechanism is accretion. Figure 7.4 shows two initially separate ice crystals. When they touch, a neck forms between them that is filled in until eventually the two crystals become one.

Both accretion and ripening of ice crystals occur in ice cream. Ripening is more important when the ice content is low, whereas accretion is generally the dominant mechanism when the ice content is high. Recrystallization leads to a deterioration in the quality of the ice cream. Figure 7.5 shows a plot of the sensory smoothness of ice cream as a function of ice crystal size. As the ice crystal size increases the texture of the ice cream becomes less smooth. When the ice crystals become very large (~100 μm) they can be individually detected in the mouth and the texture becomes icy and gritty.

Nature has provided a potential way of overcoming recrystallization. Many plants, insects and fish that live in cold climates synthesize antifreeze proteins, also known as ice structuring proteins (ISP). These molecules adsorb to the surface of ice crystals and control their growth, thereby helping to protect the organisms against freezing damage. In ice cream, ISPs slow down Ostwald ripening dramatically and also

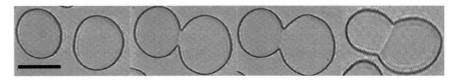

Figure 7.4   *Accretion of two ice crystals*
(Scale bar is 30 μm)

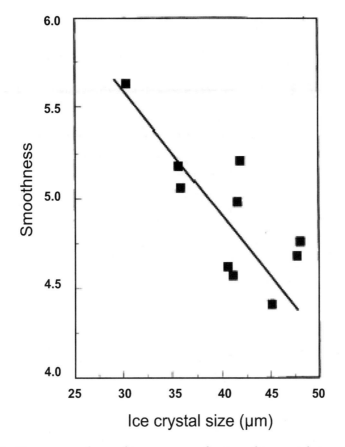

Figure 7.5 *Sensory smoothness of ice cream as a function of ice crystal size*
(Reprinted from 'Influence of Freezing Conditions on Ice Crystallisation in
Ice Cream',[1] Copyright 1999, with permission from Elsevier)

change the size and shape of ice crystals. As a result, ISPs can preserve
small ice crystals, and hence smooth textures, through distribution and
storage (Figure 7.6). They could also provide new ice structures, and
hence textures. The major challenge in exploiting ISPs is the development of sustainable, commercially viable sources capable of delivering
the quantities needed for widespread use.

The main function of ice crystals is to provide cooling. In Chapter 4
we saw that the latent heat constitutes a large part of the heat that is
removed from the mix during freezing. Similarly, the latent heat makes
up a substantial part of the heat that is removed from your body when
you eat ice cream. So if you have two ice creams at the same temperature but with different ice contents (such as the standard and soft scoop
ice creams in Table 7.1) the one with the greater ice content will provide

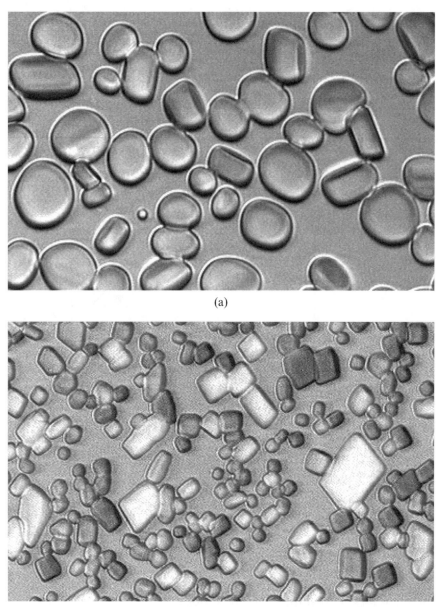

(a)

(b)

Figure 7.6 *Effect of ISP on ice crystal size and shape:* (a) *ice crystals in the absence of ISP and* (b) *in the presence of ISP*
(The images are 125 μm wide)

more cooling. This is why most people will choose an ice lolly rather than an ice cream when they really want to cool down. Eating ice cream very quickly removes lots of heat from the mouth. The body reacts to this by changing the flow of blood to the brain, which can cause an 'ice cream headache' (the medical term for this is sphenopalatine ganglioneuralgia). However, there is some debate as to whether the headache arises because the blood vessels expand in order to increase the flow of blood, to replace the lost heat, or because they constrict in order to decrease the flow and therefore reduce the amount of heat lost by the body.[2]

Ice crystals also have important effects on the rheology and mechanical properties of ice cream. However, because the other components also contribute to these, rheology and mechanical properties are discussed in the context of the whole composite material below.

Matrix

The matrix is a solution of sugars, stabilizers and milk proteins. The concentration of these solutes is significantly higher than in the mix because about 75% of the water in the mix is frozen, *i.e.* the matrix is freeze-concentrated by a factor of about 4. Freeze-concentration of the sugars has several important consequences. Firstly, we can see from the sucrose state diagram (Figure 2.10) that freeze-concentration moves the matrix closer to the glass transition. Below the glass transition temperature, the matrix becomes so viscous that it does not flow and the solute molecules cease to be mobile. Recrystallization is so slow that, on practical time scales, it effectively stops. Thus, if ice cream is stored below the glass transition temperature of the freeze-concentrated matrix, its quality should not deteriorate. Above the glass transition temperature, the rate at which changes occur depends on the difference between the storage temperature and the glass transition temperature: the closer the storage temperature is to the glass transition temperature, the slower the change. The glass transition temperature depends on the formulation, and is typically between −30 and −40 °C. In practice, it is sufficient to get close to the glass transition temperature, so factory cold stores are kept at −25 °C or below. However, it is possible to raise the glass transition temperature deliberately by using high molecular weight sugars, for example corn syrups. Whilst this makes the ice cream more stable, it can have adverse effects on the texture, because higher molecular weight sugars produce less freezing point depression, and on flavour, because corn syrups can taste unpleasant.

Another consequence of freeze-concentration is that the sugar concentration can become so high that the matrix becomes super-saturated and the sugars crystallize out of solution. This is particularly the case for lactose because it is the least soluble sugar in ice cream. Ice cream in which lactose crystals have formed is particularly unpleasant because the crystals are very hard and give it a sandy texture. Figure 7.7a is an optical micrograph of lactose crystals. They are easy to recognize because of their characteristic arrowhead shape.[3] Lactose crystallization is normally avoided by calculating the expected lactose concentration in the matrix from the formulation. If the concentration would be too high, the amount of skimmed milk powder (about 50% of which is lactose) is lowered.

Sucrose crystallization can cause another defect known as 'white-spot', which occurs most often in water ices. If water is lost from the matrix by evaporation from the surface of the water ice, the sucrose concentration rises. Eventually, sucrose can crystallize out of solution as sucrose hemiheptahydrate (*i.e.* crystals containing 3.5 water molecules for each sucrose molecule). This appears as an unpleasant-looking white spot (Figure 7.7b). Sucrose is very slow to crystallize so

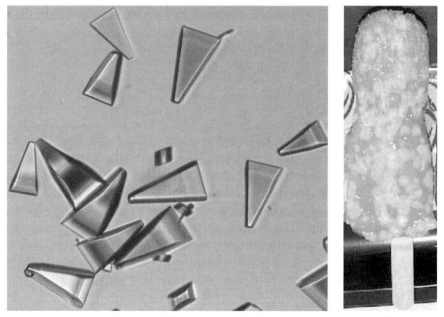

(a)   (b)

Figure 7.7   (a) *Optical micrograph of lactose crystals*
                    (The image is 330 µm wide)
                    (b) *sucrose hydrate crystallization (white spot) on a water ice*

whitespot is quite rare. It can be avoided by replacing some of the sucrose with other sugars, for example corn syrup.

The stabilizer concentration is normally below the entanglement concentration in ice cream mix, but above it in the matrix because of freeze-concentration. Stabilizers perform several functions in ice cream, most of which are related to the large increase in the viscosity of the matrix that they cause. These functions include slowing down the rate of melt, masking the perception of large ice crystals in the mouth and lubricating the ice crystals during the slush-freezing of water ices, thereby preventing the factory freezer from icing up. Figure 7.8 compares the meltdown of stabilized and unstabilzed ice creams. The stabilized ice cream loses mass more slowly. Too much stabilizer can cause an unpleasant gummy texture and make the mix so viscous that it is difficult to process. Some stabilizers slow down ice recrystallization, although the mechanism by which they do so is not fully understood and is an area of current research.

Some stabilizers also have more specific effects. LBG is especially useful in ice cream because it produces a smooth texture and a slow meltdown. It is also effective at slowing down recrystallization. LBG can form a gel when freeze-concentrated which creates a barrier around the ice crystals, thereby preventing accretion.

κ-Carrageenan is added to ice cream mixes in small amounts (about 0.02%) to prevent wheying off. Rather than remaining uniformly mixed, milk proteins and stabilizers tend to separate into separate regions of the matrix, because it is energetically favourable for them to be in contact with like molecules rather than unlike ones. κ-Carrageenan forms a weak gel that hinders the coalescence of

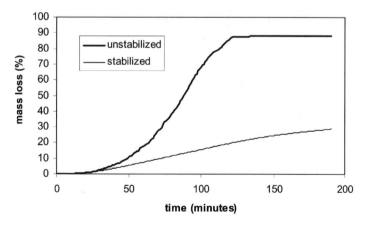

Figure 7.8 *Meltdown curves for stabilized and unstabilized ice creams*

microscopic phase-separated regions and thereby prevents separation on a larger scale.

Maras is a traditional Turkish ice cream that has a distinctive chewy and stringy texture (Figure 7.9). It is made using sahlep, a

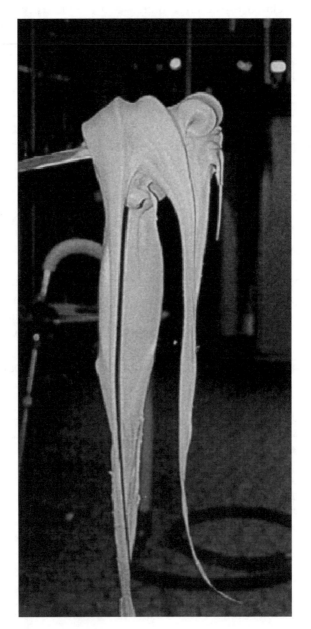

Figure 7.9   *Maras ice cream*

polysaccharide from the tubers of orchids. Its unusual rheological properties result from the interaction of the milk protein and the sahlep in the matrix. The wild orchids are rare, but fortunately a way to reproduce an elastic, extensible ice cream using conventional stabilizers has recently been developed.[4]

Fat

Figure 7.10 shows the fat structure of ice cream that is formed in the manufacturing process. Discrete and partially coalesced fat droplets are present both in the matrix and on the surface of the air bubbles. Although the fat emulsion could lower its energy by coarsening, fat is very insoluble in water so coarsening cannot take place by Ostwald ripening.

We have already discussed in Chapter 4 the key role that the emulsifier plays in creating this structure by controlling the stability of the emulsion and hence the amount of partial coalescence. Figure 7.11 shows fat particle size distributions for ice cream made with and without emulsifier. The ice cream made without emulsifier shows a single peak at a size of about 1 μm, corresponding to the small fat droplets formed during homogenization. When emulsifier is included some of the fat is destabilized. This appears as a second peak at about 10 μm.

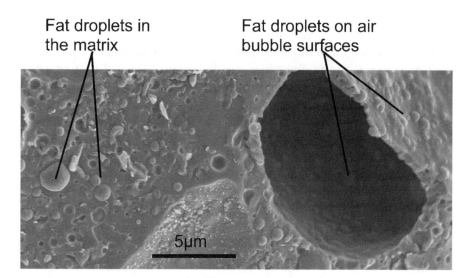

Figure 7.10 *Scanning electron micrograph showing fat droplets in the matrix and on the surface of the air bubbles*

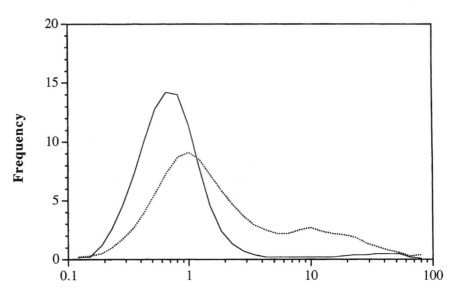

Particle Size (μm)

Figure 7.11 *Fat particle size distributions in ice cream made with (---) and without*
(—) emulsifier, showing different amounts of fat droplet coalescence
(Reprinted from 'A Study of Fat and Air Structures in Ice Cream',[5]
Copyright 1999, with permission from Elsevier)

Figure 7.12 shows a cluster of partially coalesced fat droplets on the
surface of an air bubble. Typically, 30% of the fat is partially coalesced,
but this can vary substantially.

The nature of the emulsifier affects partial coalescence. The hydro-
carbon tail of mono-/diglycerides can be either saturated (*e.g.* glycerol
monostearate) or unsaturated (*e.g.* glycerol monooleate). Unsaturated
emulsifiers produce greater partial coalescence because they displace
more protein from the fat droplet surface and so the interface is
weaker.

The most important role of the fat in ice cream is to stabilize the air
bubbles. This is discussed in the next section. Fat also has several other
functions. Firstly, it has a major effect on the sensory properties such as
thickness and mouth-coating. This is why premium ice creams contain
more fat than standard ones. Secondly, some flavour molecules are
soluble in oil but not in water. Thirdly, the solid fat particles increase
the matrix viscosity, and hence reduce the rate of meltdown. The
greater the fat content and the extent of partial coalescence, the slower
the meltdown. Figure 7.13 shows meltdown curves for ice creams with

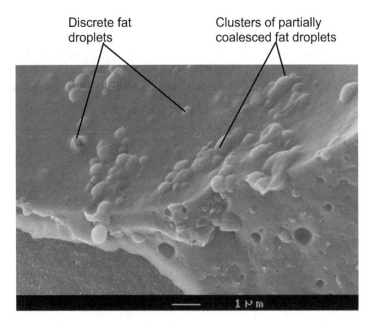

Figure 7.12 *Scanning electron micrograph showing a cluster of partially coalesced fat droplets and separate fat droplets*

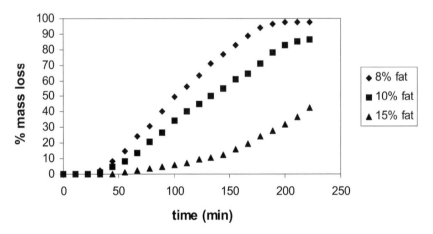

Figure 7.13 *Meltdown curves for ice creams with fat contents of 8, 12 and 15%*

fat contents of 8, 12 and 15%. The 8% sample has the fastest meltdown, and the 15% sample melts the slowest.

Making low fat ice cream is difficult because you have to find other ways of performing the fat's various functions. Polysaccharides can help to stabilize the air bubbles and increase the matrix viscosity, but can impair the texture when used at high levels. Microscopic particles of protein (*e.g.* whey protein or egg white) or carbohydrates (*e.g.* microcrystalline cellulose) can be used to mimic the fat droplets. However, these cannot entirely reproduce the characteristic mouth-feel of fat, which arises because it melts during consumption. Furthermore, other means must be found of delivering oil-soluble flavours. An alternative way to reduce the calorific contribution of fat is to replace it with edible but nondigestible lipids such as sucrose fatty acid polyesters. However, these can have a waxy mouth-feel and some have an undesirable laxative effect.

Air

Figure 7.14 shows air bubble size distributions at three different stages on leaving the factory freezer, at the end of hardening and after being thermally abused by being held at $-10\,°C$ for five days. The mean air bubble size is initially 23 μm. The dispersion of small air bubbles (like other dispersions) has an inherent tendency to coarsen. After hardening the distribution is broader and the mean size is 43 μm. (This is very similar to the increase in size on hardening in a different sample that was shown in Figure 4.16.) On abuse, this increases to 84 μm and the distribution becomes very broad, with a small number of crystals larger than 100 μm.

The two coarsening mechanisms for air bubbles are coalescence and disproportionation. Coalescence takes place when two bubbles come into close contact and the film between them ruptures, as with fat

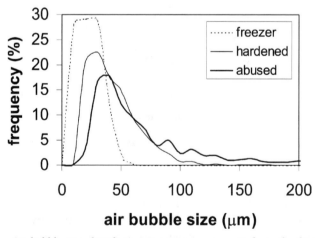

Figure 7.14   *Air bubble size distribution in ice cream on exit from the factory freezer, after hardening and after being abused*

droplets. Disproportionation occurs because the pressure inside a bubble is larger than that outside it. The difference, Δp, balances the surface tension, γ, and depends on the radius of the bubble, r.

$$\Delta p = \frac{2\gamma}{r} \qquad (7.2)$$

Thus the smaller the bubble, the larger the pressure. According to Henry's Law the solubility of a gas is proportional to its pressure. Therefore in a dispersion of bubbles of different sizes, the air in the small bubbles is more soluble than in the large ones (*cf.* the higher solubility of small ice crystals in equation 7.1). The dissolved gas can diffuse through the matrix and come out of solution in the large bubbles. Thus, there is a net transfer of air from small to large bubbles, ultimately leading to the disappearance of the small bubbles and an increase in the mean size. Disproportionation of air bubbles is analogous to Ostwald ripening of ice crystals.

There are two mechanisms by which the air bubbles in ice cream are stabilized against coarsening. The first arises from the adsorption of proteins at the air bubble surface, which lowers the surface tension. The hydrophilic part of the protein lies in the matrix and the hydrophobic part at the air surface. This reduces the driving force for disproportionation and hence slows it down but cannot eliminate it altogether. Adsorbed proteins also reduce coalescence by steric stabilization. The second stabilization mechanism is due to the adsorption of fat droplets at the air bubble surface. We can see how this works by looking at the air bubbles in a related system, whipped cream shown in Figure 7.15a. Whipping cream has a much higher fat content (about 35%) than ice cream. Thus there is sufficient fat for the whole surface to be covered with small fat droplets. These fat droplets are coated with milk proteins. The ratio of fat to protein is higher than in ice cream, so the protein layer is thinner. It is weak enough for the droplets to partially coalesce during whipping without the addition of emulsifier. The partially coalesced fat droplets form a three-dimensional structure that stabilizes the air bubbles, and links them together to form a foam. This could not happen if the fat were liquid, because the fat droplets would simply form large spherical fat globules when they coalesce. The presence of solid fat is therefore essential to stabilize the foam. This is why cookery books say that you should chill the cream and bowl before whipping. If the cream gets too warm, the solid fat melts and the partially coalesced fat structure cannot be formed. Experiment 3 in Chapter 8 demonstrates this.

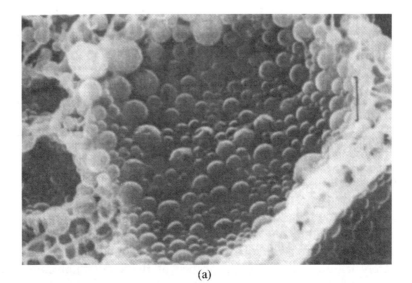

(a)

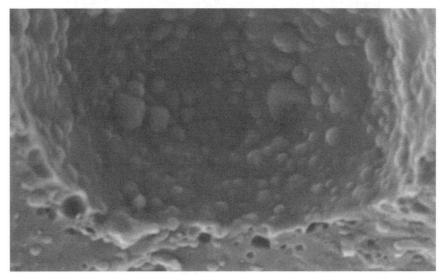

(b)

Figure 7.15 *SEM images of an air bubble in* (a) *whipped cream and* (b) *ice cream (image width 13.5 μm)*
[Part (a): Reproduced with permission from 'The influence of Emulsion Stability on the Properties of Ice Cream', I.J. Campbell and B.M.C. Pekin, IOF Special Issue no. 9803, pp. 25–36. (International Dairy Federation – Diament Building, 80 Boulevard Auguste Reyers, 1030 Brussels, Belgium, e-mail: info@fitidf.org, http://www.fitidf.org)]

In ice cream the fat content is lower, so there is not enough fat to cover the whole surface of the air bubbles (Figure 7.15b). The discrete and partially coalesced fat droplets are somewhat hydrophobic because they are partly coated with emulsifier. As a result they adsorb at the air bubble surface. They stabilize the air bubbles by forming a barrier between them. They also increase the matrix viscosity (since they are suspended solid particles), which strengthens the films of matrix between the bubbles and hinders coalescence. The extent to which the partially coalesced fat in ice cream exists as discrete clusters or an extended network (as in whipped cream) is an area of current research.

Just as the ice crystals make ice cream sensitive to changes in temperature, the air bubbles make it pressure sensitive. Normally the pressure does not change much after ice cream leaves the factory freezer. However, there are some situations in which it does experience severe pressure changes, for example at very high altitudes where atmospheric pressure is low. Boyle's law states that the product of pressure and volume is a constant for a fixed mass of gas at constant temperature. Therefore, when the pressure decreases, the volume increases. The pressure at an altitude of 3000 m is approximately 70% of the pressure at sea level. Thus, according to Boyle's Law, the volume of each air bubble increases by about 40%. This has two consequences: firstly the whole ice cream expands, and could, for example, force the lid off a tub. Secondly, the air bubbles expand and rupture the matrix layers that separate them, leading to coalescence. Eventually, after many bubbles have coalesced, a continuous channel forms through the ice cream. Channelling leads to an irreversible loss of air and collapse of the structure so that a tub of ice cream appears to be under-filled. In practice ice cream can survive up to about 3000 m, which is fortunate since it is popular in mountain-top cafes in skiing and walking regions such as the Alps. The highest altitude at which the author is aware of ice cream being sold is 2970 m on the Schilthorn in Switzerland. Ice cream can be made more resistant to pressure changes by making the matrix stronger, for example by adding a stabilizer that gels.

The main role of the air is to make ice cream soft. If you made ice cream without any air it would be hard, like an ice lolly. Figure 7.16 shows force–displacement curves from bend testing for four ice cream samples that are identical except that they contain different volume fractions of air, from 17 to 50%, i.e. 20% to 100% overrun. All the curves have the same shape, but the initial slope, the maximum force and the corresponding displacement decrease as the volume fraction of air is increased, *i.e.* increasing the overrun makes the ice cream softer.

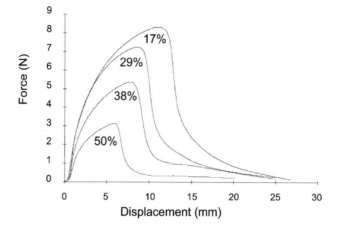

Figure 7.16 *Force–displacement curves measured by bend testing for ice cream with air volume fractions from 17% to 50%*
(Reprinted with permission from IOP Publishing Ltd.[6])

Air also has other functions in ice cream. Air bubbles scatter light and therefore affect the colour and appearance. This is why aerated ice cream is whiter than unaerated mix. The air bubble size affects the sensory properties; for example, small air bubbles give a smooth texture. Finally, the air bubbles can help to keep the ice crystals separate, and therefore reduce accretion.

So far we have only considered air. However, it is possible to use other food safe gases in the manufacture of ice cream. Liquid nitrogen can be used to freeze and aerate a mix simultaneously. The nitrogen boils vigorously, forming bubbles of nitrogen gas in the ice cream. Experiment 12 in Chapter 8 demonstrates this. Other suitable gases include nitrous oxide (which is sometimes used as an aerosol propellant for cream) and carbon dioxide (which makes carbonated drinks fizzy). One difficulty with nitrous oxide and carbon dioxide is that they are much more soluble than air, so disproportionation is faster. While it might seem tempting, you should not try to make a fizzy water ice by simply putting a bottle of a carbonated soft drink in your freezer. When ice forms carbon dioxide is forced out of solution into the headspace at the top of the bottle. The pressure inside builds up, and the bottle will eventually explode! Scientists have, however, found a way to make fizzy ice by freezing water and carbon dioxide under high pressure. An unusual form of ice is formed (known as a clathrate hydrate) in which pockets of carbon dioxide are trapped inside the crystal lattice.[7] This can be mixed with a flavoured syrup to make a fizzy water ice product.

ICE CREAM AS A COMPOSITE MATERIAL

Many of the mechanical and thermal properties can best be understood by thinking of ice cream as a composite material. The properties of composite materials are generally intermediate between the properties of the individual components. For example, the thermal conductivity of ice cream with 100% overrun is typically $0.3\ \mathrm{W\,m^{-1}\,K^{-1}}$, which lies between the values for ice ($2.2\ \mathrm{W\,m^{-1}\,K^{-1}}$), matrix ($0.4\ \mathrm{W\,m^{-1}\,K^{-1}}$) and air ($0.024\ \mathrm{W\,m^{-1}\,K^{-1}}$).

Figure 7.17 shows the Young's modulus of some water ices, *i.e.* ice–matrix composites, as a function of the ice volume fraction. The moduli are in the range 10^6–$10^8\,\mathrm{MPa}$, between the moduli of pure ice ($2.5 \times 10^9\,\mathrm{Pa}$) and matrix ($1 \times 10^3\,\mathrm{Pa}$). As you would expect, the modulus becomes larger as the ice volume fraction increases. However, it does not increase steadily: there is a sharp increase at about 60% ice. At this point (known as the percolation threshold) the ice volume fraction is so high that the crystals become connected to each other, and there is an increase in the contiguity. This results in an abrupt change in properties such as the Young's modulus and thermal conductivity, because these are determined not only by the amount of ice but also by how it is arranged, *i.e.* the microstructure and, in particular, the connectivity. The ice content is normally below the percolation threshold in ice cream

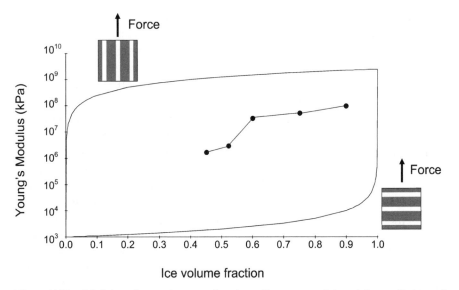

Figure 7.17 *Modulus of water ices as a function of ice content (•) and the predictions of the rule of mixtures for continuous and discontinuous ice microstructures (lines) (see text)*

but above it in water ice. This accounts for the large difference in their mechanical properties.

Figure 7.17 also shows two lines that represent the extremes in connectivity, *i.e.* where the ice is continuous (the upper line) and discontinuous (the lower line) in the direction of the force. When the ice is continuous, the modulus of the composite is close to that of pure ice even at low ice volume fractions and is given by

$$Y_{composite} = V_{ice} Y_{ice} + V_{matrix} Y_{matrix} \tag{7.3}$$

where Y is the Young's modulus and V is the volume fraction. Conversely, when the ice is discontinuous the modulus of the composite is close to that of the matrix even at quite high ice volume fractions. In this case the modulus is given by

$$\frac{1}{Y_{composite}} = \frac{V_{ice}}{Y_{ice}} + \frac{V_{matrix}}{Y_{matrix}} \tag{7.4}$$

Equations 7.3 and 7.4 are known in materials science as the rule of mixtures. In practice, the ice phase in most water ices is quite connected, but not totally continuous in any direction. This is reflected in the fact that the moduli of water ices that are above the percolation threshold in Figure 7.17 lie closer to the upper line than to the lower.

Fracture provides another demonstration of the effect of the microstructure on the macroscopic properties. When a sample is deformed, for example in the three-point bend test, the force increases until a crack forms at the top surface. This crack travels through the ice cream until it reaches the other side, at which point the sample breaks. The crack takes the easiest route, *i.e.* it goes through the weakest parts of the microstructure. The path that the crack takes therefore depends on the properties and spatial distribution of the components.

Figure 7.18 shows schematic crack paths through (a) water ice and (b) ice cream. The weakest part of the water ice microstructure is the ice–matrix interface, so the crack follows a path along the edges of the ice crystals as far as possible. In the ice cream, the air bubbles are the weakest links so the crack passes through them. The ice crystals are the most fracture-resistant components so in both cases the crack passes around them. Ice cream is effectively a particulate-reinforced composite material, *i.e.* a material in which tough particles are embedded in a relatively soft continuous phase in order to increase the overall fracture resistance. Since the microstructure is heterogeneous, it is not possible to predict exactly where the crack will form or precisely what path it

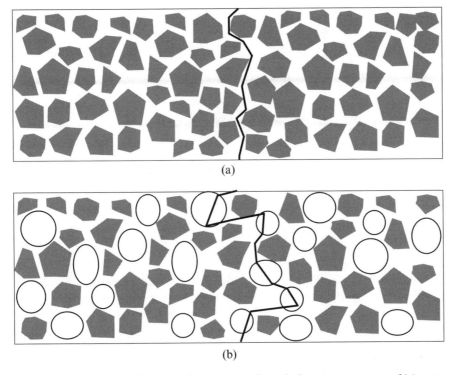

Figure 7.18   *Schematic diagrams of crack paths through the microstructures of* (a) *water ice and* (b) *ice cream*

will take, so there is an element of randomness in the fracture process. This is why it is usual to perform a number of repeat measurements on identical samples and calculate the mean values of the mechanical properties.

The crack path determines the manner in which the sample deforms and fractures (known as the failure mechanism), and hence the shape of the force–displacement curve. Although the initial slopes and maximum forces of the curves in Figure 7.16 change as the air volume fraction is increased, all the curves have the same shape, *i.e.* the failure mechanism is the same. In contrast, Figure 7.19 shows two force–displacement curves that have different shapes, which indicates that the failure mechanisms are different. In this case the samples are ice creams with different ice contents. One sample contains 50% air, 2% fat, 34% ice and 14% matrix by volume. The other has the same air and fat contents, but contains less ice (26%) and more matrix (22%). The curve from the high ice content sample has a steeper initial slope, *i.e.* a larger Young's modulus, and a steep decrease in the force after the maximum, which results from the rapid passage of the crack along a direct path through

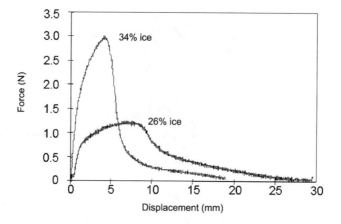

Figure 7.19   *Force–displacement curves for two ice creams with different ice contents*

the microstructure. The sample with the low ice content has a larger plastic region and a less steep decrease after the maximum, *i.e.* it fails more gently. This is because the crack passes more slowly through the sample along a less direct route. The differences between these failure mechanisms are reflected in the texture of the ice cream: the first is firmer than the second.

The ice crystals in both samples in Figure 7.19 are discrete particles. However, when a continuous ice microstructure is fractured cracks cannot avoid the ice and must pass through it. This makes the sample more resistant to deformation and harder to break. Figure 7.20 shows

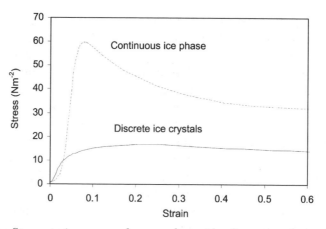

Figure 7.20   *Stress–strain   curves   for   samples   with   discrete   and   continuous   ice   microstructures*

stress–strain curves from the compression test for two samples. One has discrete ice crystals and the other has a percolated, continuous ice microstructure. The shapes of the curves are very different. As the sample with discrete ice crystals is compressed the stress increases gently. When the yield stress is reached, the sample begins to flow and as the strain is increased further the stress remains approximately constant. In contrast, when the ice-continuous sample is compressed the stress increases rapidly. It reaches a maximum at the yield stress, at which point the sample fractures and the stress decreases. This is characteristic of a brittle material.

Slush and quiescently frozen water ices provide yet another example of how the microstructure affects the properties of a composite material. Figure 7.21 shows that slush frozen water ices contain numerous small, pebble-shaped ice crystals, whereas quiescently frozen ones contain large, elongated ice crystals. Even though the formulations, and hence the ice and matrix volume fractions, are very similar, the two processes produce very different microstructures. The microstructures behave differently when deformed, and produce very different mechanical and sensory properties. The large, aligned ice crystals in the quiescently frozen water ice lock together to produce a hard product, whereas the smaller crystals in the slush frozen water ice result in a softer texture. Experiment 13 in Chapter 8 demonstrates this.

ICE CREAM AS A COMPLEX FLUID

The rheology of ice cream is very complex: it depends on the number, size and shape of the suspended ice, fat and air particles, the concentration of the sugars, proteins and polysaccharides, and the temperature. Other, less obvious factors can also have an effect, for example, cocoa particles can make chocolate-flavour mixes more viscous than vanilla ones. Most of these factors change significantly during the manufacturing process. The ice, fat and air particles are created, the concentration of the solution is increased and the temperature is decreased. As a result the viscosity of ice cream increases by several orders of magnitude. Just as with the mechanical and thermal properties, the rheology of ice cream cannot be explained by considering the components in isolation. The composite material approach described in the previous section can also be usefully applied to the rheological properties. However, the term 'complex fluid' is normally used rather than 'composite material' in the context of liquid-like phenomena.

Ice cream mix, which is a solution of sugars and stabilizers and a suspension of fat droplets, is a shear-thinning liquid. Its viscosity (η) is given as a function of shear rate ($\dot{\gamma}$) by

(a)

(b)

Figure 7.21   *Scanning electron micrographs of* (a) *slush frozen and* (b) *quiescently frozen water ices*

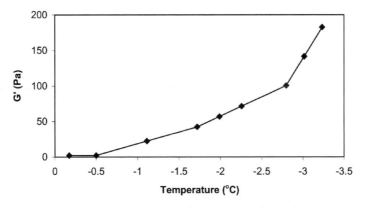

Figure 7.22 *G′ of ice cream as a function of temperature during freezing*
(Data from ref. 8)

$$\eta = b(\dot{\gamma})^n \tag{7.5}$$

where *b* and *n* are empirical constants. Fluids that obey this equation are known as power law fluids, because the viscosity varies as the shear rate raised to the power of *n*.

As the mix is frozen and aerated its viscosity increases both because the temperature is decreased and also because ice crystals and air bubbles are formed. The viscosity of an aerated mix increases according to Einstein's equation (equation 2.11) for low air volume fractions. Figure 7.22 shows what happens as an aerated ice cream mix is frozen. The storage modulus, *G′* starts to rise at about −0.5 °C when the first ice crystals form, and continues to increase as the temperature decreases. This data was obtained by using a scraped surface heat exchanger that is simultaneously a rheometer: *G′* was calculated from the torque on the dasher.

At −5 °C (the temperature at which it normally leaves the factory freezer) ice cream is a viscoelastic, shear-thinning fluid. Like the mix, it obeys the power-law equation, but with different values of *b* and *n*. As its temperature is lowered, it becomes more solid-like. Below about −12 °C it displays a yield stress whose value increases as the temperature decreases further.

MICROSTRUCTURE BREAKDOWN DURING CONSUMPTION

So far we have considered how the microstructure relates to the physical properties of ice cream. The sensory properties, however, are determined not only by the microstructure, but also by how it breaks down during consumption. Many composite materials are designed to be as strong as possible. However, ice cream must not be so strong that it does not break down because its ultimate purpose is to be eaten. Thus the microstructure must be sufficiently stable that it can be stored for a reasonable period of time, but not so stable that it does not break down easily as it is consumed. The delicate balance between stability during storage and breakdown during consumption is why ice cream has a kinetically trapped metastable structure rather than a thermodynamically stable one.

The balance between stability and breakdown is controlled by the temperature. At −18 °C or below, ice cream is stable for months, but as the temperature is raised it begins to break down. Figure 7.23 shows how the microstructure of ice cream changes as it is warmed up from −18 through −12 and −6 to 0 °C. The ice crystals melt and the matrix becomes less viscous, as a result of which the air bubbles become spherical. They also become larger due to coarsening and expansion of the air as the temperature rises.

When ice cream is eaten it is not only warmed up but also deformed by the mouth and tongue. The nature of the deformation depends on the type of product. Ice lollies are usually sucked, ice cream is spooned or bitten and then allowed to warm up in the mouth before being squashed between the tongue and the roof of the mouth, and soft ice

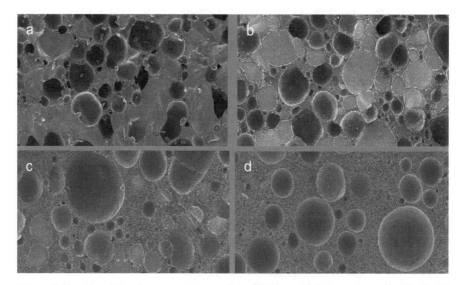

Figure 7.23 *Scanning electron micrographs of ice cream as the temperature is raised from (a) −18 through (b) −12, and (c) −6 to (d) 0 °C*

cream (*e.g. Mr Whippy*) is licked. The manner of consumption affects the rate at which the microstructure breaks down. This in turn affects the sensory experience. For example, two ice creams that have the same ice and air content but with different fat structures or different amounts of stabilizer will break down at different rates as they are eaten. The one that breaks down faster will initially have a larger surface area in the mouth. As a result, heat is removed from the mouth more rapidly, so the ice cream will initially be perceived to be colder. The other will last longer. Flavour perception can also be influenced by the rate of microstructural breakdown. Rapid breakdown gives a sudden and intense flavour burst, whereas slow breakdown produces a less intense but longer-lasting flavour.

LINKING MICROSTRUCTURE TO TEXTURE

The hardest of the links shown in Figure 1.1 to make is the one between microstructure and texture. It requires consideration of the microstructure, the physical properties, the manner in which ice cream is eaten and how the microstructure breaks down as it warms up and is manipulated in the mouth. A further complication is that while the deformations applied to samples in physical property measurements are designed to be simple, the deformation applied in the mouth is complex. For example, when ice cream is squashed between the tongue and the roof of the mouth as it melts, the deformation is a combination of stretching (Figure 6.10a) and simple shear (Figure 6.10b). Thus the sensory experience relates to more than one physical property.

Despite these difficulties, connections can be made between the microstructure and the physical and sensory properties. We have already discussed some of these, for example, the correlation between ice crystal size and sensory smoothness shown in Figure 7.5. Analysis of physical and sensory data by PCA and other statistical methods is an important tool; for example, we saw in Figure 6.18 how changes in formulation and storage conditions (which alter the microstructure) affect the sensory attributes. Another example is the relationship between the ice and air microstructure, the thermal conductivity and the perception of coldness in the mouth. The high thermal conductivity of water ice means that heat is rapidly removed from your mouth when you eat the product. This is one of the reasons why an ice lolly feels colder than an ice cream even if they are actually at the same temperature. Table 7.3 summarizes established links between microstructure, physical and sensory properties.

Table 7.3 *Links between microstructure, physical and sensory properties*

Microstructure	Physical property	Sensory property
Ice content and contiguity, air content, matrix viscosity	Young's modulus, hardness, yield stress	Firmness on spooning/ in mouth, crumbliness, scoopability
Ice content and contiguity	Latent heat, thermal conductivity	Coldness, amount of cooling
Ice crystal size, air bubble size		Smoothness, coarseness, iciness
Air content	Young's modulus	Lightness
Fat content, fat partial coalescence, matrix viscosity	Meltdown rate	Rate of melt
Matrix volume fraction, stabilizer and fat content	Viscosity when melting	Thickness, mouth coating, gumminess
Lactose crystals		Sandiness

SUMMARY

The microstructure of ice crystals, air bubbles, fat droplets and matrix is central to the physical and hence sensory properties of ice cream. Considering the components separately is sufficient for some properties, but for many others it is impossible to treat one component in isolation from the rest. Many mechanical, thermal and rheological properties depend on the whole microstructure; this necessitates a materials science approach. Furthermore, the texture experienced during consumption depends on the manner in which the product is eaten, and the way in which the microstructure breaks down. Only when all these factors are combined is it possible for the ice cream scientist to link the ingredients and the process through the microstructure to the texture. However, the current understanding of these links is far from complete and this remains an active area of research. This chapter is only a short overview of a very complex subject, and interested readers are referred to the Further Reading for more detailed treatments.

REFERENCES

1. A.B. Russell, P.E. Cheney and S.D. Wantling, *J. Food Eng.*, 1999, **39**, 179.
2. J. Hulihan, *Br. Med. J.*, 1997, **314**, 1364; J.W. Sleigh, *Br. Med. J.*, 1997, **315**, 609.

3. A. Sztehlo, *Microsc. and Anal.*, September 1994, 7.
4. A. Daniel, I.T. Norton, L.O. Lundin, R. Sutton and T.J. Foster, 'UK Patent GB 2357954', 2001.
5. H.D. Goff, E. Verespej and A.K. Smith, *Int. Dairy J.*, 1999, **9**, 817.
6. C.J. Clarke, *Phys. Educ.*, 2003, **38**, 248.
7. R.D. Bee, 'European patent EP0201143', 1986.
8. B.E. Chavez-Montes, L. Chopin and E. Schaer, in 'Proceedings of the 2nd International Ice Cream Symposium', ed H.D. Goff and B.W. Thorp, 2004, International Dairy Federation, Brussels, p. 250.

FURTHER READING

C.J. Clarke, *Educ. Chem.*, 2003, **40**, 104.
H.D. Goff, *Int. Dairy J.*, 1997, **7**, 363.
H.D. Goff, *Curr. Opin. Colloid Sci.*, 2002, **7**, 432.

Chapter 8

Experiments with Ice Cream and Ice Cream Products

This chapter describes a number of experiments on or relating to ice cream. The first seven illustrate some of the physical phenomena that have been discussed in the preceding chapters. Experiment 8 gives two recipes for ice cream mix and Experiments 9–12 demonstrate four different methods of freezing, aeration and agitation to turn the mix into ice cream. Experiment 13 describes how to make water ices. The final four describe some methods of evaluating the results. Most of the experiments use only standard kitchen equipment, but a few need more specialized laboratory apparatus. In carrying out these experiments the appropriate safety precautions must be taken, especially when using glass, heat or liquid nitrogen. Making ice cream in a classroom or your kitchen not only allows you to test the underlying scientific principles, but also to eat the results, provided of course that the ice cream has been prepared hygienically.

EXPERIMENT 1
MECHANICAL REFRIGERATION

Aim:
To show how your freezer works.

Apparatus:
A balloon

Procedure:
Your freezer works by compressing a gas so that it liquefies. The gas warms up as it is compressed, and this heat is lost to the outside *via* the tubes at the back. Once the liquefied gas has cooled down to room

166

temperature it is allowed to expand, and in doing so takes in heat from the freezer compartment. You can demonstrate this effect with an uninflated balloon. First hold the balloon at either end and feel its temperature by touching it to your upper lip. Then stretch it as far as possible, and touch it to your lip again. You should find that it has warmed up. This is because you put energy into the balloon when you stretch it. Keep it stretched for about 20 s, to allow it to return to room temperature, and then let it relax quickly. Immediately touch it to your lip again – it will now feel cool. When the balloon contracts, it does the opposite of what it does when you stretch it, *i.e.* it cools down. Stretching the balloon is analagous to compressing the gas in the freezer.

EXPERIMENT 2
STABILIZING EMULSIONS AND FOAMS

Aim:
To show how surfactants stabilize emulsions and foams.

Apparatus:
Oil (any cooking oil)
Water
A container that can be tightly sealed (*e.g.* a jam jar)
Washing up liquid

Procedure:
Put approximately 100 ml oil and 200 ml water in the container jar. Notice that the oil and water form separate layers, with the oil on top. Shake the container vigorously for a few seconds to form an emulsion, and then leave it to stand. The oil and water separate as soon as you stop shaking, and rapidly form two layers. Now add a few drops of washing up liquid and repeat. The washing up liquid contains surface active molecules that stabilize the emulsion so that it does not separate out.

To demonstrate the same effect for a foam, repeat this experiment without the oil (if you use the same container you must rinse it thoroughly first to remove any traces of the washing up liquid). Put approximately 200 ml of water in the container and shake vigorously. Bubbles form but burst very quickly when you stop shaking. When you add the washing up liquid, the surface active molecules stabilize the bubbles and a much longer lasting foam is formed.

EXPERIMENT 3
WHIPPING CREAM

Aim:
To show the effect of solid and liquid fat on foam stabilization.

Apparatus:
Whipping cream
A liquid oil (*e.g.* any cooking oil)
3 bowls
Whisk

Procedure:
Separate the whipping cream into three portions (A, B and C), each in a separate bowl. Keep A at room temperature. Add a teaspoon of liquid oil to B, and then chill both B and C (and the whisk) in the freezer for about 15 min. Then whip each of them and compare how easy it is to create a stable foam.

We saw in Chapter 4 that ice cream mix must contain a certain amount of solid crystalline fat to create a stable foam. The same is true for whipping cream. Keeping the cream at room temperature (A) or adding liquid oil (B) reduces the proportion of solid fat present and makes them harder to whip than C. Some air can be whipped into A and B because the milk protein can also stabilize air bubbles.

Whipping (churning) cream at warm temperatures leads to complete, rather than partial, coalescence. This results in the formation of large fat particles and eventually produces a fat-continuous material. This is how butter is made, and is the reason why excess fat coalescence during ice cream manufacture is known as 'buttering'.

EXPERIMENT 4
FREEZING POINT DEPRESSION

Aim:
To show that solutes lower the freezing point.

Apparatus:
Water
Salt
Ice cube tray
Scales
Domestic freezer
Thermometer (optional; digital thermometers are convenient)

Procedure:

Make salt solutions with a range of concentrations (*e.g.* 0, 5, 10, 15, 20%) by weighing out appropriate amounts of salt and water (for example a 5% solution can be made from 1 g of salt and 19 g of water). Gentle heating will help to dissolve the salt in the most concentrated solutions. Fill a couple of holes in the ice cube tray with each solution and place it in the freezer. Look at the tray from time to time (approximately every 15 min) and note when ice begins to form in each section. Measure the temperature if you have a suitable thermometer. To get an accurate value for the freezing point you need to measure the temperature soon after ice has started to form. If you measure it before ice has formed, the solution will be supercooled (see Experiment 5 below), and if you measure it when a substantial amount of ice has already formed the freezing point will be further depressed by freeze-concentration.

The higher the salt concentration, the longer the solution takes to freeze and the lower the temperature at which it freezes. You should find (unless your freezer is colder than most) that for the highest concentrations the freezing point is depressed so much that the solutions do not freeze. If you have measured the freezing points, you can plot them on the phase diagram for salt solutions (Figure 2.9). Your results should lie close to the line D–E.

EXPERIMENT 5
SUPERCOOLING AND NUCLEATION

Aim:

To demonstrate supercooling and nucleation.

Apparatus:

Water
Salt
Beaker
Scales
Domestic freezer
Thermometer (optional)

Procedure:

Nucleation is a difficult phenomenon to control, so it may take more than one attempt to get this experiment to work. Make approximately 200 g of a 15% salt solution, and place it in the container inside the freezer. Leave it for several hours so that it cools down to the temperature of the freezer. The solution should be supercooled – *i.e.*

below its equilibrium freezing point, but still liquid because nucleation has not occurred. (If it freezes, your freezer is colder than average, and you should try again with a higher salt concentration.) Now drop a small piece of ice into the solution; frost from the inside of the freezer is suitable. The ice crystal provides a nucleus for crystallization so the solution freezes. (If it does not freeze when you drop in the ice crystal, the temperature is not low enough: try again with a lower salt concentration.) The next experiment also demonstrates nucleation.

EXPERIMENT 6
SUPERSATURATION, NUCLEATION
AND LATENT HEAT

Aim:
To demonstrate supersaturation, nucleation and latent heat.

Apparatus:
A reusable hand warmer of the kind that you recharge by placing in boiling water, available from outdoor activity shops
Thermometer (optional)

Procedure:
The hand-warmer contains a supersaturated solution of sodium acetate. (Supersaturation is very similar to supercooling, except that it is the solute, rather than the solvent, that crystallizes when nucleation occurs.) When the heat is required, you click a disc on the hand warmer, which causes nucleation, initiating crystallization of sodium acetate. This releases the latent heat and warms your hands. The hand warmer is reusable because you can re-dissolve all the crystals by heating it, *e.g.* by placing it in boiling water. As it cools down to room temperature, the solution becomes supersaturated again and, provided no crystals remain, it can be kept in this state until it is needed again.

EXPERIMENT 7
VISCOSITY

Aim:
To investigate the rheological properties of some Newtonian and non-Newtonian fluids.

Apparatus:
Water
Sugar

Corn flour or custard powder
Plastic funnel and stand to hold it
Beaker
Small flat card
Stopwatch

Tomato ketchup
Jar
A weight, such as a large nut (the metal sort, not the edible one!), AA battery or other object of similar size and weight
String

Measuring jug
Large bowl or tub
Food colouring (optional)

Procedure:
Sugar and polymer solutions. Make a concentrated sugar solution (about 70% by weight) by dissolving the sugar in water (heating gently will make it easier to dissolve the sugar). Next mix 20 g corn flour (or custard powder) in 20 ml of water in a saucepan and then add 400 ml of boiling water while stirring. The heat releases starch (a polymer) from the granules in the corn flour so the solution thickens.

Make two marks on the funnel, one near the top and one about half-way down. Place the funnel in the stand above the beaker (Figure 8.1). Hold the card against the bottom of the funnel and fill the funnel with water to the top mark. Measure the time it takes for the level to fall to the second mark when you remove the card. Refill, repeat the measurement several times and calculate the mean time. Then repeat with the sugar solution and the polymer solution. The time is a measure of the viscosity.

Ketchup. Tie the weight to the string and lower it into the empty jar. With the weight touching the bottom of the jar and the string taut, make a mark on the string level with the top. Remove the weight and fill the jar with tomato ketchup to about 1 cm below the rim. Stir the ketchup well, then quickly place the weight on the surface of the ketchup and release the string. Measure the time it takes to reach the bottom, *i.e.* when the mark on the string is at the top of the jar. Remove and clean the weight, and leave the ketchup to stand for 15 min before repeating the procedure.

You will find that the weight takes longer the second time. This is because tomato ketchup is a shear-thinning fluid, *i.e.* its viscosity decreases when it has been sheared. When it is left to stand, the

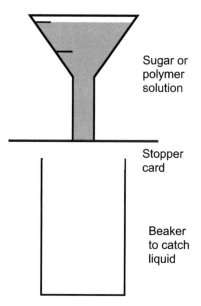

Sugar or
polymer
solution

Stopper
card

Beaker
to catch
liquid

Figure 8.1   *Apparatus for measuring the relative viscosity of liquids*

fibres from the ground-up tomatoes come into contact and form a 3D structure. This gives the ketchup a high viscosity. Stirring disrupts the structure and the viscosity decreases (*cf.* xanthan gum). This is why you have to shake or tap the bottle to make the ketchup flow. While this behaviour may be undesirable when you want to pour ketchup out of a bottle, it is useful in other fluids. For example paint should flow easily when being applied, but should then stay in place and not run down the wall.

Concentrated suspensions. Warning: this experiment can be quite messy! Measure out two volumes of custard powder (or corn flour) and one volume of water. Put the custard powder in the bowl (adding a few drops of food colouring for visual effect if you wish). Gradually add the water and mix with your fingers until all of the custard powder is wet and the mixture becomes a liquid. Stop adding water when you can tap the surface without any splashing. It does not take very much water to change the consistency dramatically. If you overshoot, add a bit more custard powder. You will know when you have the right balance because the mixture will behave very oddly.

This mixture is shear-thickening. At low shear rates, for example when gently poured or stirred, it flows easily and your fingers can move freely through it. However, at high shear rates it behaves like a solid. When you try to stir it quickly the spoon gets stuck and cracks appear.

Try pouring from one bowl into another, pinching it between your fingers, dropping some from a spoon back into the bowl, rolling it into a ball and throwing it against a wall (outside) or punching a large bowl of the mixture.

Unlike the cooked corn flour in the first part of this experiment, the starch polymer remains trapped in the starch granules and the fluid is a concentrated suspension rather than a polymer solution. When the suspension is sheared gently a layer of water between the starch granules lubricates their flow past each other. However, fast deformation forces the water out from between the granules so there is much greater friction between them and they cannot flow. Quicksand behaves in a similar manner. Ice cream mixes, the matrix and ice cream itself are all shear-thinning; if these were shear-thickening like the concentrated custard powder suspension, ice cream would be impossible to process, and would become solid every time you tried to chew it!

EXPERIMENT 8
MAKING ICE CREAM MIX

Aim:
To make ice cream mixes for freezing in Experiments 9-12. The first is quick and easy. The second is more complicated, takes about 30 min to prepare and requires constant attention, but aerates more easily and makes better ice cream.

Mix 1

Ingredients:
284 ml/half a pint of milk
284 ml/half a pint of whipping cream
3 tablespoons icing sugar
3 drops vanilla essence

Procedure:
Simply mix all the ingredients together. An easy way to make chocolate or strawberry mix is to use flavoured milk instead of the milk and vanilla essence.

Mix 2

Ingredients:
284 ml/half a pint of double cream
284 ml/half a pint of single cream

3 egg yolks
2 tablespoons caster sugar
3 drops vanilla essence or other flavouring

Procedure:
Whip the double cream until it forms peaks and then place it in the fridge to chill. It is best to whip it in short bursts: if you whip it for too long it will turn into butter (see Experiment 3). Next beat the egg yolks, vanilla essence and sugar together. Then heat the single cream up to boiling point in a saucepan and pour it onto the egg mixture while stirring. Heat gently and whisk it until the mixture thickens enough to coat the back of a spoon. It is best to use a double saucepan, or a bowl over a saucepan of simmering water rather than direct heat. If you over-cook it, the ice cream will have a poorer texture. Do not scrape any overdone bits off the pan into the mix. Pour into a large bowl and allow to cool. When it is cold, fold in the chilled whipped cream.

There are many variations on these recipes. 'Ices: The Definitive Guide'[1] contains an excellent selection, and also takes a more scientific approach than most cookbooks. You can also make other flavours, for example using mushed fruit. It is fun to experiment by creating your own unusual flavour combinations. Try for example strawberry, white chocolate and black pepper, or if you are feeling adventurous, savoury ice creams. Mustard ice cream is excellent (provided it is not too sweet) while banana, parmesan and garlic makes a surprisingly good combination!

EXPERIMENT 9
FREEZING WITH ICE AND SALT

This experiment is reproduced from ref. 2 with permission from IOP Publishing Ltd.

Aim:
To show how a mixture of ice and salt can be used as a refrigerant (this effect was the main method of refrigeration for ice cream making in the 19th century).

Apparatus:
250 ml mix from Experiment 8, or another recipe
2 litres crushed ice
8 tablespoons (120 ml) salt
1 large Zip lock bag (large enough to hold all the ice comfortably)

1 small Zip lock bag (to hold the mix)
Hand towel or gloves to keep fingers from freezing
Thermometer (optional)

Procedure:

Mix the ingredients together in the small Zip lock bag and seal tightly (very important!), only allowing a small amount of air to remain in the bag: too much can force the bag open during shaking. Put half of the ice in the large bag and mix in half of the salt. Then put the bag containing the mix on top of the ice. Finally, add the rest of the ice and salt on top, and seal the large bag, making sure to remove all the air. The ice and salt should get very cold (if you have a suitable thermometer you can measure the temperature), so wrap the bag in a towel (or use gloves) and shake vigorously for several minutes. A more exciting but riskier alternative is to wrap the bags up in newspaper, and tape well with parcel tape to form a ball. You then throw this around for several minutes – preferably outside. You can minimize the risk of salt and ice leaking into the ice cream by double-bagging the mix.

By mixing the correct proportions of salt and ice, the temperature will get down to about $-20\,°C$ (the eutectic temperature). This freezes the mixture into ice cream, while shaking stops the ice crystals becoming too large and mixes in some air so that the ice cream is not too hard. It is difficult to achieve a high overrun using this method. The best results are achieved with Mix 2 since it is already aerated.

EXPERIMENT 10
FREEZING ICE CREAM IN YOUR FREEZER

Aim:
To make ice cream at home without an ice cream maker.

Apparatus:
Ice cream mix, *e.g.* from Experiment 8
Hand whisk (preferably electric)
Freezer
Plastic tub

Procedure:
Place the mix in the plastic tub in your freezer. After an hour, remove it and whisk it thoroughly to break up the ice crystals, before putting it back in the freezer. Repeat one hour later, and then leave it in the freezer until it is completely frozen.

EXPERIMENT 11
DOMESTIC ICE CREAM MAKERS

Aim:

To see a scraped surface heat exchanger in operation. Domestic ice cream makers work on the same principle as factory freezers, but since they do not operate under pressure you can see what happens inside.

Apparatus:

Ice cream mix, *e.g.* from Experiment 8
Domestic ice cream maker

Procedure:

Simply make the ice cream according to the ice cream maker's instructions, and then compare the results with the other methods, as described in Experiments 14–16 below.

EXPERIMENT 12
MAKING ICE CREAM WITH LIQUID NITROGEN

Equipment:

Ice cream mix, *e.g.* from Experiment 8
An equal volume of liquid nitrogen
Large stainless steel mixing bowl (not glass or plastic which could break when they get very cold)
Large wooden or plastic spoon
Personal protective equipment

Procedure:

Liquid nitrogen is extremely cold ($-196\,^{\circ}\text{C}$). You should only do this experiment if you are competent to handle liquid nitrogen, and have complied with all relevant safety procedures, such as wearing the appropriate personal protective equipment. Any spectators should be far enough away that they are out of range of any splashes. You should ensure that the liquid nitrogen is not exposed to the atmosphere for a long time before use to avoid condensation of volatiles in it.

Place the mix in the bowl and pour in half of the liquid nitrogen, being careful not to splash. Stir with the spoon. Clouds of fog will emerge from the bowl. Then add the rest of the liquid nitrogen and continue to stir until the ice cream is stiff. This takes about 1 min for a litre of mix. Allow any excess liquid nitrogen to boil off before serving.

The liquid nitrogen is very cold so that ice crystallization is nucleation dominated and a large number of small ice crystals are formed. As the nitrogen boils vigorously it forms bubbles of nitrogen gas in the mix so there is no need to pre-aerate the mix. This is similar to aerating with air because about 80% of air is nitrogen. This method freezes and foams the mix very rapidly, and has in fact been used to set the world record for the fastest ice cream ever made.[3]

EXPERIMENT 13
SLUSH AND QUIESCENTLY FROZEN WATER ICES

Aim:
To compare the texture of water ices made from the same ingredients but frozen in different ways.

Apparatus:
100 ml water
25 g caster sugar
Juice of 1 lemon
Two ice lolly moulds and sticks
Whisk, or domestic ice cream maker

Procedure:
Make a water ice mix by dissolving the sugar in the water and adding the lemon juice (or simply use fruit juice). Fill one mould with the mix, insert a stick and place it in the freezer until it is completely frozen. Then partially freeze the rest of the mix using the method of Experiment 10 or 11. When the mix contains a large amount of ice, but is not completely frozen, fill the second mould, insert the stick and place in the freezer until it is completely frozen.

Compare the texture of the first (quiescently frozen) and second (slush frozen) lollies. The slush frozen lolly should be noticeably softer and have much smaller ice crystals. The ice crystals in the quiescently frozen lolly are larger and visible to the naked eye. Alternatively, compare the textures of a *Callipo* and the red core of a *Twister*, purchased from a shop. These are made from similar recipes, but the former is quiescently frozen, whereas the latter is slush frozen.

EXPERIMENT 14
MEASURING OVERRUN

Aim:
To measure the overrun of ice cream.

Apparatus:
Ice cream mix, *e.g.* from Experiment 8
Ice cream, made from the mix, *e.g.* by Experiments 9–12

Measuring jug
Tongs
Balance capable of reading to 1 g
Large container of chilled water

Procedure:
The expression for calculating the overrun (equation 4.5) can be rewritten as

$$\text{overrun} = \left(\frac{V_{ic} \times \rho_{mix}}{m_1} - 1 \right) \times 100 \tag{8.1}$$

First, determine the density of the unaerated mix [by weighing a known volume (density, ρ_{mix} = mass/volume]. If you are using ice cream bought from a shop let some of it melt and measure its density, being careful to ensure that no air bubbles remain.

  If you have a suitable volume displacement chamber you can use the method described in Chapter 4 to determine the density of the ice cream. Alternatively use the following method, based on Archimedes' principle. Weigh the piece of ice cream (m_1) and the container of chilled water (m_2). Then, with the container still on the balance, fully immerse the ice cream just below the surface of the water using the tongs, being careful not to touch the sides of the container (Figure 8.2). Since the ice cream is less dense than water, a downward force is required to hold the ice cream under the surface (the upthrust). As a result the new reading (m_3) will be larger. The difference between the weight of the container and water before and after immersion of the ice cream is equal to the upthrust, which, by Archimedes' principle, is equal to the weight of water displaced, *i.e.*

$$(m_3 - m_2) \times g = \rho_w \times V_{ic} \times g \tag{8.2}$$

where ρ_w is the density of water (1.00 g cm^{-3}) g is the acceleration due to gravity and V_{ic} is the volume of the ice cream. Rearranging equation 8.2 gives

$$V_{ic} = \frac{(m_3 - m_2)}{\rho_w} \tag{8.3}$$

from which the overrun can be calculated using equation 8.1.

  A typical factory made ice cream has an overrun of 100%. Home made ice cream often has a lower overrun since the method of aeration is less sophisticated.

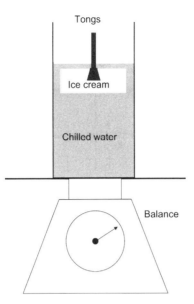

Figure 8.2 *Apparatus for determining the overrun of ice cream using Archimedes' principle*

EXPERIMENT 15
MEASURING HARDNESS

Aim:
To investigate the effect of overrun, temperature *etc.* on the hardness of ice cream.

Apparatus:
Ice cream, either made by Experiments 9–12 above or purchased from a shop
Drinking straw
Large nail (must fit inside the straw)
Thermometer

Procedure:
Hold the straw vertically just above the ice cream and drop the nail through it onto the ice cream (Figure 8.3). Remove the straw and measure the depth to which the nail has penetrated the ice cream. Then move the straw a few millimetres and repeat. The straw ensures that the nail falls vertically from the same height each time. Do this several times for each sample and take the average penetration depth. The greater the penetration depth, the lower the hardness.

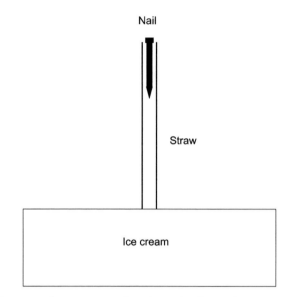

Figure 8.3 *Apparatus for measuring the relative hardness of ice cream*

Repeat using ice creams with different overruns, or use the same ice cream but let it warm up between sets of measurements. Plot a graph of the penetration depth against overrun (which you can measure using the method of Experiment 14) or temperature.

EXPERIMENT 16
SENSORY EVALUATION

Aim:
To compare the sensory properties of different ice creams.

Apparatus:
Two or more different ice creams either made by the methods described above or purchased from a shop.

Procedure:
Evaluate the samples for some of the sensory parameters discussed in Chapter 6. It is easiest to focus on one or two parameters at a time, and compare two samples. If you compare a premium ice cream and a cheap one, the differences should be clear. You could also compare ice cream made from the two mixes in Experiment 8, or fresh ice cream with ice cream that has been kept in the freezer for a long time. See how the sensory measurements of lightness and firmness compare with measurements of overrun and hardness from Experiments 14 and 15.

EXPERIMENT 17
EFFECT OF TEMPERATURE ON FLAVOUR
INTENSITY

Aim:

To show that the intensity of a flavour depends on the temperature of consumption.

Apparatus:

Two identical ice lollies

Procedure:

Allow one ice lolly to melt and collect the liquid. Eat some of the frozen lolly, then drink some of the melted one and compare their flavour and sweetness. The liquid is more strongly flavoured and sweeter than the lolly and you will probably find it too sticky and sweet to drink.

REFERENCES

1. C. Liddell and R. Weir, 'Ices: The Definitive Guide', Grub Street, London, 1995.
2. C.J. Clarke, *Phys. Educ.*, 2003, **38**, 248.
3. 'The Guinness World Records 2002', Guinness World Records, London, 2002.

FURTHER READING

Further experiments on food and cooking (not just ice cream) are suggested in P. Barham 'The Science of Cooking', Springer, Heidelberg, 2001 and N. Kurti and H. This-Benckhard, *Sci. Am.*, 1994, **270**, 66.

Index